The Gamification of Work

FOCUS SERIES

The Gamification of Work

The Use of Games in the Workplace

Emmanuelle Savignac

iSTE

WILEY

First published 2016 in Great Britain and the United States by ISTE Ltd and John Wiley & Sons, Inc.

ISTE Ltd
27-37 St George's Road
London SW19 4EU
UK

www.iste.co.uk

John Wiley & Sons, Inc.
111 River Street
Hoboken, NJ 07030
USA

www.wiley.com

Library of Congress Control Number: 2016958348

British Library Cataloguing-in-Publication Data
A CIP record for this book is available from the British Library
ISSN 2051-2481 (Print)
ISSN 2051-249X (Online)
ISBN 978-1-78630-123-9

MIX
Paper from responsible sources
FSC® C013604

Contents

Introduction

Journey to the Heart of the Gamification of Work

The precursor to what we today call "gamification" within work organizations, the first so-called "business" game (game of competitive simulation), began in Switzerland in 1926, in what was called the "house of fictional commerce" of Galliker. What we will later call *business games* developed in the 1950s. According to Kaufmann *et al.* [KAU 76, p. 17], the first business games, designed to train senior managers, actually appeared in 1956–1957, "from the efforts of the American Management Association (AMA) and of the Mac Kinsey Co.". Little by little other forms of game emerged, whose field expanded from purely commercial simulations to include team building, management training, recruitment and evaluation of staff or their emulation.

But one of the (surprising) avant-gardes of management games and the gamification of work was Soviet Russia. The American researcher Mark J. Nelson [NEL 12] reports Lenin's valorization of "socialist competition", a principle which Stalin would take up under the term "socialist emulation" (Nelson indicates that this lexical change signaled the wish to not put the workers in competition but to push them to do their best). With this principle, performance was encouraged in mills and factories with the aid of points and medals (for example the Order of the Red Banner of Labor). These competitions did not allow or award bonuses or material gains (considered too reminiscent of capitalist principles) but to displays "of encouragement and recognition". Nelson reports that in Soviet Russia, the aim of the game was not to encourage productivity alone, but that also sometimes, games might be organized around "the elevation of the cultural level of the worker" or sporting contests. He finally stresses the "mandatory" dimension of such games whose goal was to stimulate productivity and whose participants, although coerced, were supposed to "voluntarily" achieve "ever higher production quotas" [NEL 12, p. 26].

Turning toward serious games and in a direction which we might call more "mechanistic", of transferring game activities into work activities in the form of games that are no longer directly social but mediated by machines, again the process has relatively ancient roots if we consider the genesis of "Learning Machines" reported by Bordeleau [BOR 99]. The first patent for one of these machines – which were not, he explains, originally games – was filed in 1809 in the United States by H. Chard for a mode of teaching reading involving two rolling strips of paper. There followed an attempt by Edison at a Home Teaching Machine based on the phonograph. In the 1920s, an American psychology professor named Pressey proposed a Drum Tutor (1924). This machine was presented as an automated quiz that functioned by the validation of successive stages. The following decades saw educational schemes using radio, television and subsequently computer programming. Bordeleau identifies the formulation of a ludic principle linked to these machines as early as the 1950s with the cybernetician Gordon Pask, who argued that "a sort of dialogue must be established, a sort of co-operative game between the student and the machine, which must adapt to the student's answers and not the reverse. The machine must take account equally of the student's poor responses and their good responses, of the type of error made and of the response time; it must vary the difficulty of the questions based on these data" [BOR 99, p. 13]. It was finally in the 1970s that the first conceptualization of the serious game appeared in its current use and definition [ABT 70]. In effect, Abt proposed designing simulation games for teaching purposes, initially not exclusively in computing, although he had himself, as Alvarez reports, "worked on the design of TEMPER, a computer simulation game used for Cold War training" [ALV 12, p. 94]. For Abt, all kinds of games can be included among serious games, whether company role-playing games or even outdoor games [ALV 12]. Their primary goal is not amusement, as he explains in the introduction to his work, but education: "We are concerned with serious games in the sense that these games have an explicit and carefully thought-out educational purpose and are not intended to be played primarily for amusement" [ABT 70, p. 9].

I.1. Ludification and managerial practices in the "fun work environment"

Concerning the increase in the use of games in enterprises, it seems to coincide in France with the new modes of management appearing in the last quarter of the 20th Century; a phenomenon linked to the democratization of higher education and gradually individualized demands for more freedom, creativity and authenticity in work relations [BOL 99]. Demands for emancipation, pleasure in work, and creativity in ways of performing one's own work were raised by more and more qualified individuals, and could now be mediated by games and the dimensions of pleasure in play. These uses of games would lead to what would be called the "ludification" of the universe of work, ensuring a festive, convivial environment within which games would have their place. By ludification we do not intend what would be a French translation of "gamification", but a wider meaning as suggested by Bonenfant and Genvo [BON 14] and Picard [PIC 08] referring to the increasing importance of the ludic (incorporating celebration, leisure, games, media, etc.) in society. Ludification leads to the "more general trend where games (not only video games, which are certainly a factor, but games in general, the ludic) take a more and more important place in today's society" ([PIC 09] cited by [GEN 14]). The traditional concept of a game which must take place in a space and time separated from those of work has been gradually supplanted by a thought where the distinction between the two categories play/work is no longer so clear, and according to which a ludic, festive dimension may be useful to work, as permitting knowledge-sharing, relaxation, and motivation. The time and space of work have become ludified. If communication within some organizations has led to the development of areas or times for play in order to relax employees (flipper tables, ping pong, go-kart racing, network games within the company, etc.), we thus speak of ludification.

The main objective of ludification in the immediate context of work is "fun" or amusement, which prevails here over the question of learning. These initiatives would be observed in the United States, then in France, as early as the 1990s, beginning with the so-called "new economy" sector (digital start-ups, Internet, video games) and in entertainment[1]. What North American human resources departments define as a "fun work environment" would consist of: "a fun work environment intentionally encourages, initiates and supports a variety of enjoyable and pleasurable activities that positively impact the attitude and productivity of individuals and groups (...) a fun work setting is created through actions, including funny, humorous or playful activities, that publicly communicate management's belief to the employee that the personal and the professional accomplishment he or she has achieved are valued by the organization" [FOR 03, pp. 22–23]. Nelson speaks of "funsultants", meaning management consultants who promote ludification: that is, working toward the development of a work environment that combines celebrations, games and relaxation. The breakdown of the traditional work/leisure distinction, underlined by Boltanski and Chiapello [BOL 99], will thus take place in a professional universe that has become compatible with personal fulfillment, amusement and relaxation among employees, in the aim of efficiency at work.

Keeping employees amused at work by providing them with space for games or sport, or time for celebration, is therefore part of this ludification, and its direct objective is team bonding and supposed "well-being at work", which is argued to foster the creativity and confidence conducive to taking initiative [SAV 03]. Indirectly, it promotes longer hours spent at work [BAL 09], a good internal as well as external image for management, and in particular attracting new employees [FOR 03] or even a reduction in absenteeism and turnover[2]. The issues for "funsultants" are motivational, and aim to act on the context of work – to build emotional ties between the

1 Note that the survey conducted by Ford *et al.* [FOR 03] among more than 500 organizations would show that these management methods would be more widely used in business than in the "non-profit" sector (NGOs, associations, administrations, etc.), and in particular games. This is the same if the employees of these businesses are less unionized; the higher the rate of unionization, the less competition between employees and the fewer festive activities take place [FOR 03, p. 26].

2 Studies by Kinnie *et al.* [KIN 00] on the practice of games in two call centers show their effectiveness in reducing turnover (moving from 25 to 10% for one and from 30 to 8% for the other). This is one of the "benefits" identified by managers interviewed in the study by Ford *et al.* [FOR 03]. We will also suggest this argument for a *reversal day* event in a hotel chain.

employee and their colleagues as partners in play and at their place of work, to leave no room for boredom, to channel stress[3] – rather than to act on their skills as such as, for example, *serious games* attempt to do. There is however a secondary goal to this motivational goal: the possibility of intervening in people's behavior at work and in particular, according to the oft-repeated formula, improving their "know-how-to-be": energy, smiling and "positive emotions" may also be spread within a business, but also to partners and customers [BAL 09, KIN 00, ALF 03, FOR 03]. Baldry and Hallier stress the insufficiency of previous managerial strategies to get workers committed, not to their work, but with respect to their organization: "(...) these efforts have failed to develop substantial levels of workforce commitment, but also employees generally have seen the contrasts between employers' messages of mutuality and the short-term, hard HRM reality" [THO 03, BAL 07]. Far from a willingness to accept management accounts, most employees have either deployed a resigned, often skeptical compliance, or they have attempted to mimic management's own "rhetorics in order to protect their positions by appearing to be 'on side'" [COL 97, HAL 04, BAL 09 pp. 14–15]. The organization of games, entertainment, and the creation of spaces devoted to relaxation, to friendliness or to leisure embodies the rapprochement, desired by management as well as by employees, between elements that could be considered in opposition: work and relaxation, hierarchy and proximity, competition and pleasure. Above all, by displaying the concern of leaders for individual well-being at work – today renamed "quality of life at work" – they center the individual and their development; an individual in not only their professional aspects but also those relating to "not-at-work".

Baldry and Hallier underline the congruence between the development of open spaces and these activities: "team competitions and fancy dress days can only be made to work in a non-hierarchical open plan work space" [BAL 09, p. 19]. In addition to the already-noted goal of communication, there is a collective dimension to the ludification of the environment that should be considered. There is a social dimension to these ludification

3 "Enthusiasm", "group bonding", "satisfaction", "creativity", "friendly working relations", "reduced anxiety and stress" and "confidence with one's colleagues" are the benefits highlighted by Ford *et al.* in their study of businesses practicing a "fun work environment". Note that these terms are not used by the employees themselves but by their managers, in reference to what they believe such a management style brings to their subordinates.

activities that is central to the motivations of the organizations putting them into place. Playing together allows not only getting to know each other but to appreciate each other, and favors a "good atmosphere" at work, supposed to encourage investment and productivity.

Gamification, ludification and ludicization

As distinct from ludification, what we will call "*gamification*" brings the structure and mechanisms of games (role-playing games, Kapla, Lego, board games, video games, etc.) not into spaces and times dedicated to leisure within work organizations, but for the purposes of carrying out work: training, education, sales, management, etc. *Gamification*, defined by Deterding *et al.* [DET 11] as "using game design elements in non-game contexts" explicitly refers to game structure. It is this game structure that is imported into contexts other than that of play, including pedagogy (quizzes, crosswords etc.), research (for the purposes of, for example, deciphering the structure of an enzyme in the framework of research against AIDS on the crowdsourcing site Fold.it, by means of a game developed by the University of Washington), civic activities such as road safety (for example the "speed camera lottery" experiment in Stockholm, where drivers obeying the speed limit go into a draw to win prizes financed from speeding tickets), or finally work. Other authors, such as Zichermann and Cunningham [ZIC 11] define it, as Bonenfant and Genvo report, "as a process consisting of using the state of mind and mechanics of game-play to solve problems and to involve users, the basic design principles of games being applied in different contexts" [BON 14]. But note that what Zichermann and Cunningham understand by "state of mind" refers, in particular, to questions of the player's engagement in the game. Games would be expected to foster, shall we say spontaneously, this engagement, which makes them interesting to use, according to these two authors, for the purposes of marketing. We are here dealing with what we could call the "mechanical" dimension of games, and those of its functions that are inherent in its structure: rewards, indicators of progress, degrees of difficulty, etc. Zichermann and Cunningham thus speak of "reward structures, positive reinforcement and subtle feedback loops alongside mechaniscs like points, badges, levels, challenges and leaderboards" [ZIC 11].

These are indeed the components of games (or some of them, and we can ponder the selection made) which are used. This same concept of gamification is found in the texts of the American game designer Jane McGonigal [MCG 11], when she proposes four elements that she argues to be common to all games: a goal, rules, a feedback system and voluntary system. Picard [PIC 08] for his part underlines the importance of storytelling and narrativity in these transfers.

We thus distinguish *gamification* from ludification along the axis of the English language distinction, largely taken up by theoreticians, between *game* and *play*, but also *ludus* and *paidia* [CAI 67]. These categories in effect embody the distinction between the structure of games, inherent for example in chess just as in game design, and "playing", or the attitude or posture, even the state of mind of the player. Attitude, state of mind, posture, may obviously be combined with a game structure or a game objective, but may also be external to and occur independently of a game structure or objective. If, for example, I amuse myself by spinning a pen balanced on my finger and trying not to let it fall, I am playing, without a game structure and also without any objective which could be called a "game", and it is probable that I get some pleasure from it. As Stéphane Chauvier [CHA 07, p. 18] argues, the game is "detachable" from the player, as opposed to the ludic attitude: "We must be careful as to the difference between playing with a rubber band and playing the rubber-band game®". *Play* is autotelic, in that the aim of the game is solely to play it, while a game thought of as something existing in and of itself – like chess, rugby or hopscotch – is heterotelic, in the sense that it leads the player to figure out and apply certain rules or objectives. The arbitrariness of play is thus reduced in a game which remains linked to "features which structure and constitute it", that is, "a structure of practice" [CHA 07, pp. 83–84]. This dichotomy between play and game thus establishes ludification and gamification as distinct practices for management which embarks upon them, the first leading to benefits derived from its autotelic dimension, such as a good atmosphere, or indirect benefits (for example, getting to know one's colleagues), and the second leading to structured practices. Introducing such ludic principles as, for example, points or game levels into the process of work, training or marketing leads to gamification.

A third term, not widely known yet, adds a third typology to this definition of concepts linked to games: ludicization. It was coined by Genvo, who defines ludicization as the process according to which "an object which was not seen as a game becomes to be perceived as such, and whereby this change of perception may also lead to changes in the meaning of the term 'game'" [BON 14]. In cases of ludic imports into business, we can thus speak of ludicization in the case of "challenges" through which the employee is led to achieve what in current management language are called "objectives"[4].

Fulfilling objectives within time constraints is a way of working with which certain businesses are very familiar: salespeople in shops or customer service workers in call centers, for example. Transforming time constraints and quotas into a game results from this somewhat ludic "framing". This latter, supposed to reduce coercion, may be accompanied by praise for the best employees, earning points that can be redeemed for rewards (consumer goods, a food basket, etc.). We could also speak of ludicization when it comes to "exercises" in training or recruitment sessions, presented today as games as in certain "simulations". The question, however, remains of the perceptions of stakeholders in the race toward the objectives which has become a game.

Researchers have until now thought of gamification according to the principles of "game design", or in direct reference to video games. We could argue for extending this concept to the formal dimensions – structures – of other games that are brought into the work situation. Thus, introducing the theatrical format as a way of working with principles of managerial operation or the organization of work could be thought of as a type of gamification, as much as the serious game that borrows its structure from video games for purposes of training professionals. It is in this broader perspective that we will conduct our study on the gamification of work.

I.2. A socioeconomic context favorable to the emergence of gamification

Some authors hold a very critical position with regard to the contemporary tendency toward gamifying work. Thus, Bonenfant and Genvo [BON 14, p. 5/7] write: "Again (…) under the guise of a game, a vision is

4 Reference is made to the use of challenges by management in Stéphane Le Lay's text [LAY 13] studying such "games" in a call center, contrasting them to games initiated and organized by employees at work.

promoted of an economic system based on accumulation, efficiency and productivity". The two authors speak of "logics (...) of rationalization of activity through the addition of the constraints inherent in 'games'". Rewards, graded objectives, feedback on so-called "progression", encouraging competition between colleagues through counting points, importance of speed and rhythm are equally elements of game design that justify *a priori* the analysis of an intensification of cadences as much as an intensification of normative control and constraints. If this is quite clear from the reading of elements borrowed from video games, we might wonder if it is also present in games not borrowing from game design, such as role-playing games, business theater or the more carnivalesque forms of role reversal. What is the interest in games brought into the working context, in the case of those that do not use the techniques of points, levels or competition? What interest for management and, consequentially, what managerial meaning do these games have? What does the game, as a game, allow, produce or engage, with respect to work and its organization that encourages its continued, even increasing, use in contemporary organizations [FOR 03, ALL 15]?

We have spoken of the context conducive to ludification, explaining the managerial turn in the 1980s and 1990s. Boltanski and Chiapello [BOL 99], analyzing managerial discourse during the 1990s, have shown how the massification of access to higher education has led to qualified workers endorsing a so-called "artistic" critique, stemming from the social movements of the 1960s, demanding more autonomy, creativity and recognition at work, or a humanization of working relations. This demand is no longer categorical but individual: to be recognized in the workplace as an individual. In consequence, there has been a growing refusal of the status of a mere operator and the goals of graduates have smoothly moved from climbing the social ladder to finding interest in their work or even becoming accomplished in it, like an artist realizing their creation. In a period of full employment and the legacy of the emancipatory political demands raised in May 1968, this demand was recognized, or we might equally say recycled by management – what Boltanski and Chiapello call "the new spirit of capitalism". Individualizing and psychologizing work relationships had the double effect of responding to the demands of qualified workers and "motivating" them – since work thus becomes comparable, in terms of a path to accomplishment, to leisure or a "passion" – all the while diminishing union collectives, through individualization linked to the subject taking charge at work. Contemporary management wields power, according to these authors, through this sleight-of-hand: the social criticism demanding recognition of the individual at work is satisfied and, in doing so, the demand is individualized, even isolated. This in turn impacts traditional social relationships at work based on the contradiction between bosses and unions, or a collective – a union – facing managers. It is this double operation coupled with new forms of organizing work (project management, flexibility, individual evaluation, "lean" processes) which might partly explain the isolation of individuals at work and the diminishing of collectives, as much as the decrease in cooperation [DEJ 98, GAU 05] observed throughout the last third of the 20th Century.

Management in the 1980s and 1990s moved wholly toward the idea of the development of individual well-being at work. Nelson points out two reasons for this managerial trend: "The first is more mercenary: some in business hope that there exist non-monetary incentives that can elicit additional labor, thereby motivating workers with things that are 'free' (such as internal competitions and points), rather than having to pay out as many monetary incentives, such as traditional performance bonuses. The second worry is that certain kinds of productivity are simply impossible to monetarily incentivize, and instead require somehow producing intrinsically motivated, happy workers" [NEL 2, p. 24].

"Fun" and "play", like "game", in relation to ludification and gamification are always – and logically – linked with the goals of work: competition and performance in primary logic are equally linked, paradoxically – it might seem – with the goal of a good atmosphere. It is no accident if these techniques emerged originally in the leisure and service industries. For the latter, many authors underline their importance in view of the quality of service delivered: "well-managing employees' emotions and moods should be considered equally critical for a hospitality business's success" [YOU 13]. As in hospitality, and in call centers, or even in amusement parks, these must be, with respect to their clients and the services offered, nothing so much as "smile factories" [VAN 91]. For some of them, such as the Disneyland park studied by Van Maneen at the beginning of the 1990s, an individual's investment for the benefit of the product being sold occurs not only through their knowing-how-to-be but also through their body – as he puts it, taller than average, slimmer than average, young, with healthy teeth "*and a chin-up, shoulder-back posture radiating the sort of good health suggestive of a recent history in sports*" [VAN 91, p. 12]. Ludification and gamification relate through the regularity of performance that they encourage, to the sporting competition dear to contemporary organizations, to bodies which perform.

A question of generation?

Authors studying the *fun work environment* insist on the generational dimension of the populations involved. If, as we shall see, gamification tends to be extensive and to move from the professional worlds of entertainment to more traditional professional domains (hospitals, consultancies, etc.), ludification and gamification took hold originally, as we have seen, among digital and multimedia start-ups in the 1990s [SAV 03], in businesses recruiting *en masse* those who were designated by different North American and later European authors as "Generation Y" or "digital natives". This expression denotes people born between 1975 and 1995 who grew up with many inventions in the domains of media, telecommunication and high tech: computers, consoles, video games, CDs, MTV, smart phones, Internet, augmented reality, etc. [PAU 01].

In our ethnography of digital start-ups in the 1990s, we noted the specificity of newcomers to the labor market and to this new sector of the digital economy, graduates gifted with a knowledge of the practices and use of computers and games not shared by their elders holding higher decision-making roles. This inversion of the poles of experience, linked with the euphoria of the Internet bubble then taking place, promoted a festive atmosphere and the encouragement of the egos of young people seen as prodigies or, at least, more pragmatically, possessors of knowledge useful to the development of new markets.

One might think that, 15 or 20 years later, this asymmetry of knowledge no longer so clearly applies. Initially stemming from managers, consultants and media, the expression of generational specificity has however gradually been introduced to the scientific literature. It is in these terms that many English-speaking or French authors, originally on questions of marketing [PAU 01], then on management (for example among the most cited: [MAR 05, EIS 05, TAP 08, PRA 09]), then today on games in work organizations [GIN 13, ALL 15], describe these qualified workers now in their thirties: the "Y-ers".

The description of the inhabitants of this generation, despite their supposed indiscipline and volatility, seems to be the answer to management's prayers: the "Y-ers" are said to seek the same engagement with work that they find in the digital world [TAP 08]. Spontaneously as well as expertly using new technologies, they are autonomous, independent, flexible, networked, opportunist, creative, and "lean" methods seem to be made for them [MAR 05]. Their supervisors are advised as follows to manage them well: to consider them as people; to coach them like teachers in their quest for new learning; to give them wide flexibility in their use of time, projects and careers; to provide them with constant feedback and to give them praise, recognition and bonuses [MAR 05]. These are generational characteristics and these last two points lead Tascott [TAS 08] to say that gamification is a solution in the workplace for this generation. Note moreover that fun, for them, is more a prerequisite than a benefit [GIN 13].

More than a generational reality, in business, the "Y-ers" would be taken to refer to a group of highly qualified and executive personnel [PRA 09] and to the demand of those personnel, as Boltanski and Chiapello recalled earlier, for more autonomy and creativity at work. Pichault and Pleyers [PIC 12], very critical of the analytical methodology arguing for a Generation Y, show in a study carried out in Belgium over more than 850 employees, not all of them qualified, that it is not so much a matter of managerial specificities as of human resources orientations that individualize and promote collaborative work, which meet a demographic cohort: "The constitutive dimensions of this ideal type are not unrelated to certain important theoretical developments in the field of human resources" [PIC 12, p. 40], they thus note. Here, it is not a matter of employees "naturally", one might say, adapted to present-day managerial logics, but more ways of thinking about management as well as contemporary demands, such as – they mention – the importance given to "individual affirmation" [PIC 12, p. 44]. Some of the supposed specificities of "Y-ers" are in reality shared as far back as the baby-boomers, such as the meaning given to work, its utility, recognition, autonomy or creativity – old and recurring topics in motivational texts. Pichault and Pleyers' study shows how, in the postwar period – and we can assume this to be linked to new forms of capitalism – employees have gradually come to think of themselves as only being able to rely on their own resources, ceasing to think of their work organization as a landmark or a stable future. However, those whom the authors of "Generation Y" call "opportunist" show a strong aspiration to cooperation, solidarity and team spirit [PIC 12, p. 46].

Rather than a Generation Y to whom managerial responses of the fun work environment and later gamification are directed, it is a managerial doxa to which we must turn to try to grasp what happens when capital is reconciled with labor... and when work becomes a game.

I.3. Managerial uses of games in organizations

We will present our research in three chapters. The first will be devoted to presenting our theoretical framework and to the choices made, in regard to our field of endeavor, among game theories. The second will be devoted to the question of games and the encounter between games and work. Considering the (rich) theorization of games and the numerous discussions over the last century on this complex concept, do the elements of our field, which we aim to describe in this section, correspond to a game? The last stage of this writing will deal with the plural dimension of games. In the relevant literature, it is a question of framing, of translation from the real to fiction. In our field, it is a question of roles, of simulations using the devices of fiction, inversion, substitution or changing places. The term "game" is thus used in its mechanical dimension: a space between several frames, of movement, the margin of action or even, might we say, an augmented reality. We will study these aspects of the game to try to describe the spaces, times, margins and actions to which it refers in work organizations.

Finally, we will consider why to use games in work organizations. Beyond fun alone, since with games it is a question of structure, of devices, of rules: what are the expectations of management concerning it? How does it operate and what are the functions linked to work and its organization that might be its own in the field of games studied?

1

Prelude: Fun, Play, Game, Ludus... A Survey of Game Theories

Several researchers have underlined two salient aspects from which we can embark upon research into games. First, they indicate the extreme complexity of elaborating a theory of games. A first generator of complexity in game theorization consists of the great difficulty of thinking of it as a global object: how to think the similarity between playing with a rubber band between one's fingers without letting it fall (see [CHA 07]), playing a game of chess or playing a role-playing game, for example?

In the second place – and the second salient aspect of game research – we have the poverty and very relative interest brought to the field in French research communities, outside the interest of children's games by educational and psychological sciences. In fact, well-behind Anglo-Saxon game studies, research on games only really developed in France during the last decade. It is strongly marked by the development of video games (videoludic practices) and new technologies, which tends to give second place to the study of non-digital games. Yet, these were investigated previously by some avant-garde (though isolated) figures in game research such as Roger Caillois in the 1950s and Gilles Brougère today. In the United States, though there exists a field entirely devoted to "game studies", these researches concern themselves, as their name indicates, primarily with games (their rules, principles and structures) and not with play (the ludic attitude). We also find an interest in games among researchers into folklore, and this diversity and partiality in game research contribute to the difficulty of envisaging it as a global object.

Historically, as starting points for reflection on games, we could cite, on the one hand, research into animal games and, on the other hand those – as we stressed, a handful – on specifically human games (without restriction to the domain of childhood alone).

1.1. Animal play, human play

Karl Groos (1861–1946), a German psychologist, wrote "The Play of Animals" [GRO 96], then later "The theory of play" in "The Play of Man" [GRO 08], where he argued – on the basis of his observations of animal play – that play functions as preparation for later life. His research would later be seen as relating to a functionalist theory: play has a function, it serves a (biological) end, a heterotelic principle that might be controversial in theories on human play. As another point of discussion, many criticisms would be made of research into animal play, dealing with the acceptance of the word "play": what a human calls "play", is it play for the animal? Is there not an anthropomorphic bias in thinking about animal behaviors in the image of human behaviors?

These questions would further continue in anthropology, where some researchers such as Hamayon [HAM 12] underlined the great variability of what might be understood as relating to "play" in the populations studied, or again, in a more general questioning, such as Geertz [GRE 80], who called for critical examination of analogies with play and game.

Beyond these controversies relating to the ontology of "play", these researches however bring to light a common element in animal play and in the activity of human play (ludic as well as artistic, moreover): behaving "as if", pretending, the "not for real"; as in the example of puppies play-fighting, fighting "not for real". Research in ethology [FAG 81, IMM 80] show that not all animals play. Practices called "play" have been identified mainly in mammals. Birds, for example, do not play (with the exception of corvids: ravens, magpies, jackdaws, etc.) and animals, according to which species they belong to, do not play the same games.

Three types of games have been categorized among animals. Ethologists speak of locomotor games and rotation games, play with objects and social play. Pierre Garrigues, a researcher in anthropological ethology, thus describes locomotor and rotation games among animals:

> "Most locomotor games are distributed in a fairly uniform manner among the various animal species: running, running in a circle, jumping, bouncing, kicking, rolling, sliding. Others are more common, or alternatively, more original. Thus, the behavior "jumping in the air" has the widest distribution. It is found among non-human primates, cetaceans, rodents, carnivores and artiodactyls (including the hippopotamus). To this repertoire, some animals add their own specialties: chasing their own tails, as in domestic dogs or minks, or even hanging upside down, as in gibbons, red pandas or ravens. Some locomotor and rotation games involving the whole body or parts of the body, like those of young chimpanzees, have become popular in descriptions made by primatologists, whether young chimpanzees repetitively climbing up and sliding down their mother's body, their acrobatics between tree branches, or improvising pirouettes while walking" [GAR 01, p. 12].

As for play with objects, one critical doubt arises (in the absence of the ability to question the animal about what it is doing) between observation/exploration by the animal of the object, use of the object as a tool, and playing with the object. As Garrigues [GAR 01] says, there is no firm line between the three activities. "In fact, at what moment does playing with an object become the discovery of a tool?" [GAR 01, p.13], he asks. Playing with objects covers different activities such as picking up, carrying, shaking, biting or pinching, pulling to pieces, throwing up and catching, throwing away as well as pushing [DES 06].

In 1976, Egan described the behavior of a cat (quoted by DES 06, p. 52]: "typically, an object begins by being sniffed at or batted with a paw. The nature of the object determines whether it will be bitten or not; furry toys are those most commonly bitten. If it is bitten, the object may be kept in the mouth, shaken and tossed (behavior that helps stun live prey), or carried (to a corner where prey could be eaten in peace, for example). For the other type of object, an initial small blow with a paw might make the object roll, in which case it will lead to squatting and pouncing (the movement being the triggering stimulus for these two behaviors) which, as for prey, has the effect of immobilizing the object". Games with marbles, jump-rope and playing with a ball (outside of the game structure present, for example, in a soccer game) could be seen as play with objects in the human setting. This latter here has a clearly ludic function. This goes even more for objects with which one prepares to play: balancing a pen on one's finger, etc.

As for the last category of play, the social, it is very distinct among animals from human play and we seem to meet here again the difference between game and play. Animal play, when social, seems to refer only to the latter category, as opposed to human games, which socially structure play: tennis, soccer, or monopoly, for example. By social play, we mean in ethology – still according to a functionalist reading – the fact that play allows members of a group to get to know each other and to be able to agree. In addition, play explores social positions (who is dominant). Play, by promoting interactions, reinforces links between the members of the group.

Social play among animals involves fighting, agility (primates sliding down their mother's body, for example), pursuit or possession, serving either biological or social purposes. Ethologists note that social play among animals is mainly a game of simulation: simulating aggression, defense or mating. Klaus Peter Köpping thus says of play that it is a "pivot" category, "linking the social and the natural" [HAM 12, p. 298].

If in these theories play has a function in the development of the young animal, this is greatly emphasized for children's games [PET 84, MIL 79, WIN 80].

> "Through the superabundant physical activity deployed, games doubtless participate in the physical development of the animal, but this is not the only benefit. In its interactions with the environment the young animal develops its social and cognitive skills. It experiments, in conditions which are relatively safe, in varied situations, in the frame of which it learns to find solutions to new problems: find the appropriate distance in interaction with its peers, or discover the use of a tool. Through its explorations, the young animal thus develops behavioral regularities with regard to the physical and social environment. From this point of view, the central function of play is to allow in the young individual the "unlocking" of different activities,

belonging to its species' repertoire or developing from gradually acquired patterns of action" [PET 84].

The function of development is one of the recurring arguments for the use of games in training. However, limiting "play" to a function of development does not work as well for the case of "games": does a game of monopoly or cards help us develop[1]?

To continue on the subject of animal play, ethologists have shown what we can call "codes of communication" that are linked to it. When animals fight "in play", they show the signs of "not for real", characteristic of play – and which approach, for many theoreticians, the game of fiction. This indication of "not for real" becomes necessary so that a playful bout does not turn into a real fight (this being true for animals as well as humans). This is what Bateson, as we will return to at greater length, calls the *metacommunication pertaining to play*. Bateson tells us that when we play, animals as well as humans, we send a message indicating: "this is play". This message is non-verbal for animals and, for humans, can be verbal ("let the games begin", "game on!") or non-verbal or even arise from the context or the accessories of the game (taking out a monopoly board puts the act of buying real estate into a different context, meeting a troll avatar means *a priori* that someone is not attacking you for real, etc.). Not knowing or being able to understand this metacommunication, that is, the figurative dimension particular to play, is a symptom of schizophrenia according to Bateson.

So it is that for animals, we remark that if dangerous tactics are used in a fight between animals, "in a game in contrast, [these tactics] and bites are absent, as well as the stereotypical signals of threat and submission. (...) the "physiognomy of the game" is always present, as an indicator superimposed on acts modeled on those of actual combat, but without the same amount of violence" [GAR 01, p. 15]. Garrigues adds that the "physiognomy of the game" [...] "is used by individuals to indicate their availability to play and prevent their partners from any misunderstanding during playful combat" [GAR 01, p. 16].

It is thus this physiognomy of the game that signals play among animals. In this sense, by using metacommunication, animals show that they are

1 The possibility of a social function in these games is of course understood; and yet, is it possible to speak of "development" without using this term in an overly broad manner?

playing. Play may sometimes be solicited by an animal using a very particular message:

> "The best known is found among canids, under the form of a "play bow", displayed by the dog to invite a peer or a human to begin or continue a session of play. Crouched on the ground, the back bent in the arc of a circle and the thorax pointing towards the partner, the dog keeps its front legs flat in front of it; he is ready to jump one way or the other. This posture is only seen in the context of play" [GAR 01, p. 16].

Ethologists thus show that, in play, there is not only intentional communication among some animals but also use of the figurative dimension.

1.2. Theories of human play

The French word *jeu*, meaning both play and game, comes etymologically from the Latin "jocus" – "joke, or play on words". Consulting historical dictionaries of the French language, it is explained that *jocus was frequently associated with ludus (play in action) and eventually absorbed its meaning.*

> "Jeu", since its first appearances in 1080, has indicated, again according to the dictionary, "free amusement" and "ludic activity in as much as it is organized by a system of rules defining success and failure, winning and losing" (1160). Its dimension of regulation led to the word applying to sporting competitions (1160) and then to the theater (1200). A century after its appearance, "jeu" also applied to battle [HAM 12].

1.2.1. *Precursors*

One of the earliest theorists in the field of games study is Johan Huizinga (1872–1945). In his 1938 work, *Homo Ludens* [LED 38], he examined the "social function" of play, to which he allocated a role for humanity equal to that of *Homo faber* (the capacity of humans for creation and work) and

Linnaeus' *Homo sapiens* (knowledge, intellectual power). Huizinga's thesis is very all encompassing (in the history of humanity, everything started as play), something which would be critiqued by those who followed up his reflections on play. He says that play was the origin of culture: that play was not born from culture but that from play, he argued, culture came.

Thus, Huizinga was led to consider "all human activity as nothing but pure play" [HUI 38: 11], in addition to which "human civilization begins and develops within play, as play" [HAM 12]. Differing from the tradition of condemnation of play by religion in Europe, his theory removes it from the solely moral register. Historically, and particularly as a simulacrum or representation ("acting as if"), play is condemned by religion – in particular, by Christians – as Roberte Hamayon reports, citing Tertullian (theologian of the second century after Christ): "Can theatrical masks please God? If he forbids the likeness of any living thing, all the more shall he forbid that anyone disfigure his image. No, no, the author of truth loves not that which is false". Tertullian again: "games trick human beings and thus betray the will of their creator". Play is assimilated to the inauthentic, to trickery, to artificiality – which is still often the case: fooling, feinting, simulating, etc. An often-quoted formula of Freud argues that "the opposite of play is not seriousness, but... reality" [FRE 88, p. 34].

Huizinga is thus one of the first to try to theorize play and to articulate all these disparate things that are called "play" outside of all moral considerations. He therefore characterizes play according to the following properties, very often repeated after him and actively supporting the debate on the concept of game:

– Play is a "free" activity: "All play is first and foremost a free action. A commanded game is no longer a game" [HUI 38, p. 24]. Play is distinguished from compulsory activity, which could be further clarified (Brougère [BRO 05] proposes replacing the criterion of freedom with that of decision) or discussed if one considers, for example, ethnographic data (Hamayon reports that among the Buryats, play is compulsory), or even a parent playing with their child, actions that are certainly often free, but sometimes coerced. In other words, guilt or resignation faced with the child's demands might function as a "command". This definitional aspect directly confronts our subject with the question of possibility of play in work organizations, if we consider that it is organized there and features neither the spontaneity nor the absence of coercion inherent in the idea of liberty.

– The goal of pleasure: "The child and the animal play because they find pleasure in playing, and their freedom lies there. (...) Play is superfluous. (...) At any moment the game may be deferred or abandoned. It is not imposed by physical need, even less by moral duty. It is not a task. It is performed during "leisure time" [HUI 38, p. 24]. One might question here the possibly anthropomorphic dimension of these remarks. Susanna Millar thus questions the criteria allowing us to assert that an animal or a young child is contented. She underlines the fact that observations of children do not show a systematic link between play and contentment [MIL 79]. Is it always a pleasure to play? When we lose? Or for oneself, when we play to give pleasure to others? Huizinga next defines play according to three parameters: the first is specific to play that operates in – we might say – an anthropologically specific space and takes a human being out of their daily activities. The two others are more classically spatiotemporal. Play is thus a practice linked to "a need for isolation" [HUI 38, p. 40].

– An activity that takes the player out of their "routine life": "Play is not life as "routine" or "in itself". It offers an excuse to escape this to enter into a provisional sphere of activity for its own sake [HUI 38, p. 24]. This idea of "a provisional sphere of activity for its own sake" consequently brings into question the use of play in work organizations, just as in pedagogical practices. Far in these cases from being a "pure" activity (play for its own sake), play is supposed to promote learning or experience, for example. It becomes hybridized with work activities.

– A temporally bounded activity and an activity circumscribed in space: "Play begins and, at a certain movement, "ends" [HUI 38, p. 26]. "The local limitation of play is even more striking than its temporal limitation" [HUI 38, p. 27]. The question arises here of the applicability of this criterion in particular to play when one is interested in the practices, observed in the sociology of work [BUR,79, ROY 58, DES 91, SHE 07, LE 13, DUJ 15], of workers transforming their activity from work to play. Does the delimitation of play apply in this case as neatly as to a game of bridge?

To these characteristics Huizinga adds the rules: "Every game has its rules" It is these which, added to the spatial and temporal limits, give the game its characteristic that "it creates order, it is order" [HUI 38, p. 27].

Finally, three last properties define every game:

– the "tension" generated by the game, or a factor of uncertainty;

– which other authors such as Bateson or Gilles Brougère would call the figurative dimension: "In an authentic game, besides its formal traits and its cheerful atmosphere, one essential characteristic is also indissolubly associated: the awareness, even if relegated to the background, "of acting only in appearance" [HUI 38, p. 43]. "Behaving as if" or "not for real" is accompanied for Huizinga by the awareness of the facticity of what is being performed. An author like Henriot will question the degree of awareness of the game by the player in the play of animals or very young children.

– Finally, according to Huizinga, a game has no purpose apart from itself. A game is "autotelic", has no goal but to be a game. We will return to this.

One of the main limitations of Huizinga's work is that he restricts games to the sole dimension of competition. He argues this in part in view of the origin of ludus: "It is of the highest importance that the simple word ludus, despite all the joy and freedom it evokes, was always used to signify the collection of Roman games, with their bloody, superstitious and slavish character" [HUI 38, p. 111]. He thus links games back to jousting or combat, arguing that: "Battle, as a function of culture, always assumes restrictive rules, demanding, up to a certain point, the recognition of a ludic quality" [HUI 38, p. 130].

This might quite obviously be debated: is a child playing with a skipping rope in competition, even if only with themselves? In role-playing games, is it always a question of competition? Of being on top? In the warrior example provided by Huizinga, is this "ludic quality" always recognized? And, if so, by whom?

The summary provided by Huizinga of the elements of every game consists of a few lines, from which further researchers would construct – either in opposition or in agreement – game theory. "A game is an action which takes place within certain limits, of place, of time and of will, in an apparent order, following rules freely consented to, and outside the sphere of utility and of material necessity. The atmosphere of a game is of rapture and enthusiasm, whether a sacred game or a simple celebration, a mystery or entertainment. The action is accompanied by feelings of transportation and tension and brings with it joy and relaxation" [HUI 38, p. 187].

In studying games, Huizinga is led to think that they arise prior to culture. Culture, according to him, has throughout human history followed games. Rules of order and limit generate culture through their structuring dimension. Huizinga speaks of "higher forms of social games" [HUI 38, p. 75]. "Social games" are understood by Huizinga as: "actions ordained by a community or group, or by two groups with respect to each other" [HUI 38]. Huizinga thus combines the functionalist and evolutionist theories previously seen in the ethological analysis. According to him: "The more the game is able to raise the level of life of the individual or group, the more truly it transforms into culture" [HUI 38, p. 76]. The game therefore has the function of development and evolution for the human group and the individuals who comprise it. Although thought of as autotelic, devoid of purpose, he says [HUI 38, p. 78], it can be seen to have some kind of a function that we might qualify as "higher", not linked to the protagonists of the game, but to the humanity to which they belong. Play would have an anthropological and social power of producing culture. This thought of the higher operationality of play is taken up again in two ideas:

– the fact that it is thought of by Play Theory as a medium, vector of "adaptive potentiality" [SUT 97, p. 229];

– the fact that it is thought of by anthropology in a very similar manner to ritual and shares with the latter, according to ethnographic observations, the property of generating "the expectation of an "effect" on a different order of reality" [HAM 12, p. 88]. Roberte Hamayon questions the prospective dimension linked to the game: "On what basis arises this expectation of a possible "effect" of the act of playing? (...) can it be perceived as capable of influence on another level than that on which it occurs? (...) It would seem that it is in the margin and through metaphorization that playing operates as a modality of action (...)" [HAM 12].

The question of effects on a dimension exterior to its own framework (production of culture, adaptation, effect on the harvests for the Buryats studied by Hamayon, etc.) is one of the recurrent themes of research on games and leads, in a way, to think of this subject as linked to another: this time, productive power.

This way of conceiving games would, paradoxically enough, be in agreement with another dominant thought in this research: play as the opposite of work and productivity, immediate this time, one of whose main thinker is Roger Caillois.

Another 20th Century author important for the study of this field, Roger Caillois, tried to propose a definition of play through a typology that goes beyond the single theme of antagonism (competition, combat) proposed by Huizinga. As with many game theoreticians who followed, Caillois (1913–1978) made the play/work opposition one of the axes of definition of what play is.

For Caillois, play has to do with entertainment, "it rests and it amuses" [CAI 67, p. 9]. According to him, play is devoid of constraints (we once again find the idea of freedom) and above all, like Huizinga, he considers it to be free from consequences in the world outside of the game or the life of the player: "creating neither property nor riches nor a new element of any sort; and, except for the transfer of property among the circle of players, leading to an identical situation to that at the beginning of the game" [CAI 67, p. 42]. In this sense, play produces nothing and aims to produce nothing. These categories explain in part the logic of opposition to work as soon as one thinks of play: play is unproductive, work is productive, play is free, work constrains, play is fun, work does not have this purpose, work is an activity with consequences, play is the opposite.

Caillois even opposes the literature proposing children's play as having an educational function: "On the contrary to what is often claimed, play is not training for work. Only in appearance does it anticipate adult activities. The boy who plays horses or trains is in no way preparing to become a rider or a mechanic" [CAI 67, p. 21] – a statement to be cautious with, if one considers, for example, the learning of socially expected gender roles mediated through toys (dolls, etc.).

Caillois widens the statement in making play not a training exercise for a particular activity (in contrast to those exact uses made of it in business: training for an evaluation interview to lead a reticent colleague to accept a mission, etc.). He says more broadly that play is preparation for life, notably in "developing all capacities to surmount obstacles or to cope with difficulties" [CAI 67, p. 21]. We see once again here the previously underlined paradox of immediate productivity denied to play in favor of a higher function of adaptation.

Another paradoxical dimension of play that we might underline is that, although free and with no productive function, it is yet not without rigor and carries within itself a principle of authority. A game is in fact composed of rules that "define what is and is not part of the game, that is, the allowed and the forbidden. These conventions are at the same time arbitrary, mandatory and without appeal. (...) nothing maintains the rule except the pleasure of playing, that is the wish to respect it. One must play the game or not play at all" [CAI 67, p. 13]. To play the game consequently means to consent to this authority.

Caillois suggests several characteristics of games in the direct tradition of Huizinga (before outlining a typology of them): In the first place, he conceives the game as an activity about which he also says that it is "free: into which the player cannot be forced without the game immediately losing its nature of attractive and joyful entertainment" [CAI 67, p. 42]. Caillois here frankly opposes the principle of pleasure and that of constraint, denying any possibility for example that work, the carrier activity of constraint, might be the source of pleasure or of "attraction" and "joy", to use the same descriptors that he uses. This sharp cleavage between activities and values underlies the recurrent concept of a definition of play as opposed to the activity of work. He then takes up the criteria of delimitation and uncertainty already developed by Huizinga: the activity of play is "separate: circumscribed in limits of space and time specified and fixed in advance", and it is "subject to conventions which suspend ordinary laws and which temporarily institute a new set of rules, the only ones that count" [CAI 67, p. 42]. Caillois also makes play a spatiotemporally autonomous sphere, but also according to the principle of rules that govern a separate space. We see how this fits with the concept of "frame" developed by Bateson and later Goffman.

It is also "uncertain: its development cannot be determined nor can the result be predicted in advance, a certain latitude for the sake of invention always being left to the initiative of the player" [CAI 67, pp. 42–43]. The complexity of play is shown here as well, in the conjunction between uncertainty and clear limits, just as previously in its capacity to combine the *a priori* scarcely compatible dimensions of freedom and regulation particular to play. Finally for Caillois, as we have said, play is an "unproductive" activity.

An interesting characterization, which will inspire many consequent reflections thereafter, is about the relationship between play and fiction.

Caillois argues that the activity of play is "fictive: accompanied by specific awareness of a secondary reality or of frank unreality with respect to everyday life" [CAI 67, p. 43]. With this category, he goes beyond the whole question of the figurative dimension or the awareness of play. With fiction, we go beyond the "not for real", that is to say, we encounter the possibility of freeing ourselves from reference to the real.

Having suggested these characteristics of play, Caillois groups games into four categories: simulation, competition, chance and dizziness:

– Mimicry (simulation games): "The game may consist (...) of a player becoming an illusory character and acting accordingly. We are thus faced with a varied series of events which have the common feature of resting on the fact that the subject plays at believing, at making themselves believe or making others believe that they are other than themselves" [CAI 67, p. 61]. Imagination, fictionality and "acting as if" are at the heart of the category of mimicry.

– Agôn (competitive games): This category refers to both individual and collective games involving the question of challenge and competition. Their goal is to overcome: oneself, another, the machine. Caillois underlines the dramatic tension inherent to the agôn: "The antagonists are applauded each time they take an advantage. Their struggle has its vicissitudes which correspond to different acts or episodes of a drama. It is finally the moment to remember the extent to which the champion and the star are interchangeable characters" [CAI 67, p. 150].

– Alea (category of games of chance): Chance, destiny and fate are at the heart of this category.

– Ilinx (games of vertigo): Turning around to get dizzy or jumping elastics are for Caillois games whose goal is the physical sensation of vertigo and euphoria.

Ilinx and Alea share a "letting go"; one submits oneself to chance or to vertigo, as opposed to Agôn or Mimicry that assume the mastery of the game or the simulation. For Caillois, these categories of game can be seen as significant of the values of certain societies that value such and such a type of game. We might note, within work organizations, the predominance of simulation and competition games, to the detriment of games of chance (forbidden, even if elements of chance may exist, for example, in the case of the allocation by drawing lots for such and such a team of competitors) and vertigo games (rare).

The absolute proscription of games of chance in organizations is perhaps linked to what Caillois mentions about Piaget on pedagogical principles, where respect for rules is taught to the child "for their moral training". In France, educators, Caillois stresses, do not promote games of chance.

For Caillois, these forms may be hybridized, but affinities or compatibilities/incompatibilities exist between the categories. He suggests notably a compatibility between, on the one hand, simulation and competition (the dramatic dimension of both competition and spectacle) and, on the other hand, vertigo and chance.

In addition to the four categories of agôn, mimicry, alea and ilinx, Caillois suggests a distinction between two categories of "ludicity": between paidia (noisiness, laughter, agitation), on the one hand, and ludus (concentration, calm, even solitude), on the other hand.

One can see that this paidia/ludus distinction comes close to other distinctions made – for example, in the English language between play and game or by Jacques Henriot between the ludic attitude and the structure of the game.

1.2.2. *Differentiation between game structure and ludic attitude*

Henriot would be the first to theorize the distinction between ludic structure (game) and ludic attitude (play). For him, "the structure expresses the schema of the action: it does not indicate its meaning. One may do something within a game; one can do the same thing without playing. It is certainly not the same thing for the subject concerned, but it is the same for someone watching the action" [HEN 89, p. 107]. The same ludic structure can, for Henriot, generate either a "serious" or a ludic activity. He takes the example of simulators, either professional or recreational (flight simulators, etc.): "The simulation is as real as the thing simulated. (...) What the pilot does in manipulating their controls is not, in essence, different to what they would do if playing. Certainly, the difference is vital: the least error on their part could cause a catastrophe. But this is not a difference at the structural level. The gestures, the procedures, the decisions, the way of thinking and acting of the operator are identical in both cases" [HEN 89, p. 56]. What makes the game is argued to be the fact of playing it. The structure might just as well support a professional activity, for example, as a leisure activity. What is meant here is that play is certainly an attitude, but equally context. One might think that a pilot might find a ludic dimension in their activity. But the question of consequences, of productivity, of constraint, etc., is here expressed in a differentiated manner.

In this sense, Henriot says that play is also to be understood as a gap – as in the "play" between two nuts on a screw – through which different uses may slip. "A toy is all the more useful 'when it leaves a play, a space of determination' within its shape and consistency as an object, allowing the child to hang imaginary frames upon it, to insert therein their own game (quotes Grange, 225)" [HEN 89, p. 94]. In any game, he says, there is "potential ludicity".

Concerning the ludic attitude, Henriot returns to his example of simulation, saying that what differentiates a game from another situation is "the intention of the actors". From here, he makes a distinction between a "fictive situation" (the game: for example, a flight simulator in a video game) and an "actual situation" (the non-game: for example, a professional flight simulator). "The difference between a 'real' and a 'fictive' situation does not emerge from the structure, but from the intention which motivates the actors, the conditions in which they operate, of the value they attribute to the goal" [HEN 89, pp. 111–112].

What is interesting here is the reference that Henriot also makes to fiction. Henriot adds, just like Freud – whom he cites[2] – a creative dimension to play. A game is in this sense a secondary space where, when one does the same thing as in reality (through simulation, for example), it is not the same thing. In the game, the real is more than mimicked, it is surpassed, with new elements added to it: actions as well as attitudes. This point is important and will be moreover stressed by consultants and trainers through games: a game is a situation which is at the same time real and factitious, and the way in

2 And his famous phrase among game theories: "the opposite of play is not seriousness, but reality", already cited.

which the players of the game simulate reality therein through role-playing is not, all the same, what would happen in reality. We are in the power of "as if", in the realm of possibility, of imagining the possibilities of action.

Henriot links this imagination to the dimension of the unexpected and uncertainty in play. In play, there is freedom of decision and accordingly, risk-taking. This allows him to explain, on the one hand, the strong link between play and the question of limits (reaching them, trying to exceed them), and, on the other hand, its equally strong link with chance, through the very fact of the uncertainty that it carries.

It is this understanding of play as a space of possibility and experimentation that we can understand with Winnicott and Bateson.

1.3. Play as potential and intermediate space

1.3.1. *Winnicott and play as "potential space"*

Winnicott (1896–1971), the celebrated psychoanalyst and specialist in child development, makes play, in his work "Playing and Reality" [WIN 71], a space of communication. According to him, play is a space neither entirely "inside" (interior to the subject) nor entirely "outside" (of the world outside the subject). It consists of an intermediary space where symbolizations are possible (and, in the development of the infant, where the first stages of symbolization take place). This space rests at the same time on the subjectivity of the individual subject and on the world outside of the subject (other people, toys, objects in the world, etc.).

Winnicott anchors this potential space initially between the baby and its mother. It is in the first place a transitional space, an "area of separation" as he puts it, born of relations of trust and affection, which will allow play to reside there. The child, increasingly detached from its mother, will be able to create there, and thereby create itself there as a subject. What Winnicott means by creativity is: "the retention throughout life of something that belongs properly to infant experience: the ability to create the world" [WIN 04, p. 55]. Play as a creative force can be questioned here as well, in its relationship to fiction: if a world is created there, how is the world recreated and what displacements operate within what is a gap?

With play, we enter the idea of a potentiality of the subject: to develop as an individual, to intervene in oneself and one's environment, to invent one's life – not only in the sense of fantasy or dreaming, but in the sense of intervening in it. Play in this sense produces effects.

Winnicott says that it is in playing, in having – we might say almost literally – introduced play between the baby and its mother (and consequently brought into existence the subject that is the child as separate from their mother) that communication becomes possible: between two subjects, but also between the interiority of the subject and the exteriority of the world.

One might consider play as a first symbolization, which says through the actions of the game that cannot yet, for the very small child, be said with words. This is why play, with Winnicott in particular, is introduced into therapy. "Playing leads into group relationships; playing can be a form of communication in psychotherapy (...)" [WIN 71, p. 41]. Play is interaction before speech. It is a space of symbolization and of communication even before the use of the major symbolic register that is verbal language when the child begins to speak.

Child psychiatrists such as Bailly [BAL 01] argue that in its first perceptions, the infant does not see itself as separate from its mother, whom they consider to make up an integral part of their being. For the psychic life of the infant, it is a question of an "illusion" that means that "internal and external realities are not yet clearly distinct for the infant, allowing "intermediate" experiences, in particularly that of possessing a transitional object that is neither the real mother, nor her internal representation, but a little of both" [BAI 01, p. 42].

This stage is one of great dependency. For specialists in early childhood, transitional objects allow the child to constitute itself as a subject by supporting separation: "The transitional object allows the child to accept the absence of the mother and gives it the possibility of having the feeling of existing despite her absences. In this way, the baby can accumulate life experiences without its mother and without finding itself in danger. The transitional object allows this game, something the child can submit to its "omnipotence", in the presence or absence of its mother (...). In this sense, Winnicott could say that the infant plays as soon as it is able to possess a non-me object" [BAI 01, p. 43]. We see here how in psychoanalysis, the idea of children's play meets that of a space conducive to the experience and development of the self. Here, we encounter previously mentioned notions developed by game theoreticians: freedom, creation, imagination, intrinsic exteriority of play (its own time and space) and finally, the perception of limits. We consider that it is through this play – space – between mother and child, that the infant grasps the boundaries of its being and the world that surrounds it. Yet, in child psychiatric literature, play is not constituted in limits but rather in space. It allows– first for the infant, testing the limits then, second for the child, breaking free from said limits. Every further game retains for Winnicott, the nature of a "transitional phenomenon". Bailly explains:

> "By "transitional phenomena" must be understood the continuity of experiences of omnipotence characteristic to children's games. When the child plays, they enter into an intermediate space, where reality no longer acts as a constraint but sees itself remodeled to suit the child's internal needs (...). The child can distinguish reality from their own desires, but play is a way of existing as "oneself", despite the constraints of reality to which it must adapt. (...) We must distinguish clearly here, as Winnicott does, a game, which can be socially organized, and the much more essential activity of playing. Playing is a creative act, the invention of an individual, which allows for an infinite number of variations, when social or educational games are much more limited. Playing is thus a transitional phenomenon. It consists (...) of a vital experience. By "vital" must be understood 'essential to the child', namely the feeling of really existing, or even the feeling that life is worth living" [BAI 01, p. 44].

Here, we see how much play is presented as an anthropological experience, fundamentally human, opening up the initial possibility (or its failure) of a self-possessing subject, capable of seeing itself in interaction with the world. Through play, says Winnicott, the objects and phenomena of the world are put into relation with the "internal or personal reality" [BAI 01, p. 105] of the individual.

For Winnicott, play characterizes psychotherapy, which endeavors to recreate a potential space between the therapist and the patient. He thus says that "In psychotherapy, what are we dealing with? With two people playing together" [BAI 01, p 84]. Constitution or reconstitution of the subject is possible through play, which is able to generate potentiality and restore to the subject its capacity to act upon itself and the world or, more simply to create or restore relations between them. Here, we find again development as a higher function recognized in play, here no longer for humanity as a species, but for the individual. More than an activity, play is action: "To play is to do", says Winnicott [WIN 71].

Winnicott suggests that in human development, there is a slippage "from transitional phenomena to play, from play to shared play and, from there, to cultural experiences" [WIN 75, p. 105].

1.3.2. *Bateson and the question of "frame"*

The concept of a link between play and communication – in its relation to symbolization and its second-degree nature – is shared by many game theoreticians, beginning with Bateson (without forgetting Henriot, who will speak of the "ludic metaphor" as soon as play is involved).

Gregory Bateson (1904–1980) was a particularly atypical researcher – anthropologist, psychologist, founder of the Palo Alto school, son of a great geneticist who had considerable influence on his relationship to science.

Bateson invites us to consider play not as a content, a "substance', but as a "form", which structures an activity. It is this form that allows us in particular to distinguish a real fight from a pretend one (not for real, in play). This form, at the moment it comes into action, presents a story about the activity that it covers. This is what he calls the metacommunicative function of play: every game is accompanied by a narrative of which the object "is here the relationship between the interlocutors" [BAT 77a, p. 248].

> "Now, this phenomenon – play – could only occur if the participant organisms were capable of some degree of metacommunication, i.e., of exchanging signals which could carry the message: 'this is play'" [BAT 72, p. 179].

If this meta-message is not conveyed, then a real fight will ensure, a fight in earnest, between knights and dragons on the playground as between two dogs.

Bateson insists on one particularly illuminating thing: the question of the play frame (which Goffman takes up and develops). He advances the idea that in the play frame there are play acts that do not mean what they would mean outside the play frame. "The playful nip denotes the bite but it does not denote what would be denoted by the bite" [BAT 72, p. 180]. The "frame" (which he names a "psychological frame") is for Bateson a collection of messages (messages which may be verbal – "one might say, you be the knight and I'll be the dragon" – or which might be actions: I imitate a ghost). It is the frame that is metacommunicative. He has an amusing formula (Bateson is often amusing) to describe it: "The frame tells the viewer that he is not to use the same sort of thinking in interpreting as the picture that he might use in interpreting the wallpaper outside the frame" [BAT 72, p. 187]. And look out for those who cannot decrypt this metacommunication and take pretend aggression for real. The conditions of communication are then threatened.

Bateson, to explain his theory of play, links the difference between play and reality to the difference between the map and the territory. He raises in the first place the question of what is preserved of the territory on the map (and the associated question of what is excluded). In a simulation game of a job interview, for example, what does it preserve of the reference situation and what does it leave out? What are the "rules of transformation", to use his expression, from one frame to the other?

With a map we do not have the ground, the territory, but a symbolization. A map is in a way a metaphor for the territory, and this goes for all games, in particular those of simulation, competition or role-playing. This is why games are often related to fiction. But a particular kind of fiction, referring to reality. A game is a balance between, on the one hand, a fictional frame, and, on the other hand, a reference frame. There is an oscillation between the two frames, which can explain why a game can quite quickly turn into a non-game (we often see this among children).

Bateson finally speaks of a "more complex" game: that which makes us doubt the nature of the game and pose the question: "is it a game?". The figurative element of which the game is a vector is not without ambiguity.

A game is often a double object: the ludic is enacted there, but also something else in reference to another frame, the reference frame, which is particularly true of simulation games that refer to a model. Its ambiguity is equally visible in what the game puts into play in terms of relations between people (and which may lead to a halt in the game when it becomes too slippery: the "time out, I'm not playing any more" of children). In the game, there may be aggression as well as cooperation. Finally, its ambivalence also lies in the fact that we play together even when we are playing against each other.

This ambiguity makes Bateson say: "In primary process, map and territory are equated; in secondary process, they can be discriminated. In play, they are both equated and discriminated" [BAT 72, p. 185].

The "primary process" may refer, in reference to adults, to a psychiatric disorder. It is on the basis of his theory of play that Bateson can notably elaborate a theory of schizophrenia.

Schizophrenics do not access the figurative dimension, and will always interpret the metaphor literally. This is why it is difficult for them to play, as in using or understanding humor. But this non-detection (anxious and anxiety-provoking) of the register being used (is it for real? not for real?) may be our own, when we wonder, for example, if someone is joking or not (in the case of dry humor, for example). In this case, we always veer toward the first degree: "This is characteristic of anyone who feels "on the spot", as demonstrated by the careful literal replies of a witness on the stand in a court trial" [BAT 72, p. 209].

A game thus involves a particular register of communication; a figurative element which, if not detected, may end the game or raise the question: "Is this play/a game?"

One might consider this question as central to the use of games in business: am I participating in a game, or a performance evaluation? Is it a game if what I do will have an adverse impact on the way my colleagues perceive me? etc. Bateson insists moreover on the importance of context: of what frames the interaction that might help someone determine whether it is a game... or not.

1.3.3. *Goffman's analysis of frame*

Erving Goffman (1922–1982), sociologist and linguist, re-examined Bateson's concept of frame. He emphasized an interest in the "putting in parentheses" allowed by the game, and the possible confusions between reality and play. For Goffman, frame is what gives meaning to the interaction. This supports the organization of events, such as a game, which organizes the scene of interaction and its meaning: "I assume that definitions of a situation are built up in accordance with principles of organization which governs events – at least social ones – and our subjective involvement in them; frame is the word I use to refer to such of these basic elements as I am able to identify" [GOF 1974, p. 10].

In his major work, *Frame Analysis: An Essay on the Organization of Experience*, Goffman defines, in greater depth than Bateson, what a frame is. For Goffman, each situation requires us to ask the implicit question: "What is happening here?". A frame indicates two things:

– it gives its definition to the situation – thus allowing us to interpret it: we thus return to the game, and to the question which it previously raised: is it a game or not?

– and consequently, it indicates the appropriate ways in which to act in this situation. If this is a game, I can allow myself to behave in ways that would be perceived as eccentric or even inappropriate in any other frame; disguise myself as an alien in a role-playing game – though respecting the indications (equivalent to rules) given regarding the context of the game. If this is not a game, but an ethnology course where I am asked as a student to consider my daily environment as if I knew nothing about it, like an extra-terrestrial, I will thus adjust my behavior and leave, for my experimental process, my alien costume in the cloakroom.

One may therefore: question the case of a simulation game putting one colleague in the situation of evaluating another. The two will not act in the same way in the game as if it were a matter of real evaluation for their work, and of their working relationship in reality. This is linked to the definition of the game as an *artifact*, a point which we will develop later in our research. At the same time, in this example, two frames will be considered at the same time: the game frame, inserted into the professional frame:

– One frame is thus constituted of implicit social rules that we integrate throughout our existence: how to speak, how much distance to leave, what posture to adopt, what categories of meaning to use etc. A suit will be just as socially inappropriate on the beach in August as a swimwear in a bank office at the same time, and the very process of our enculturation marks out our daily experience with indications as implicit as they are multiple.

– “The individuals I know don’t invent the world of chess when they sit down to play, or the stock market when they buy some shares, or the pedestrian traffic when they move through the street. Whatever the idiosyncrasies of their own motives and interpretations, they must gear their participation into what is available by way of standard doings and standard reasons for these doings” [GOF 74, p. 236].

– Clearly the question arises, in the context of games organized by the work organization, of which frame to refer to in priority. What arrangement is in place to make the two frames compatible or, at least, capable of alignment? In a way, one might consider this question in terms of a *double bind*, as conceptualized by Bateson. One might consequently think of it as a paradoxical injunction – to demand one thing and its opposite – the injunction to work-play. Now, we know from Bateson that a double bind always returns the protagonist to the level of the first frame, the frame of reference, in the case of our example: work.

Goffman talks, as Bateson did before him, of primary and secondary frames. The primary frame correlates with the activity of interpretation linked to a situation. The primary frame, he says, “allows us, in a given situation, to give meaning to such and such of its aspects, which would otherwise be devoid of meaning” [GOF 74]. The primary frame, when it is social (and not natural, Goffman tells us), returns us to norms. It is these that, once acquired, allow us to make sense of the situation and comply with its implicit laws.

As for the secondary frames, these are keyed or fabricated on the basis of primary frames. Concerning fabrication, Goffman gives the example of falsification, where the primary frame is mimicked for the purposes of deception. Deception uses the conventions of the primary frame to function as a secondary frame. In the example of games, conventions are here used for explicit purposes – there is furthermore no secondary frame (game) if their transformation is not acknowledged. Consequently, he argues that there is use and “processing” of the frames: what he calls the “keying”.

Keying, according to Goffman, corresponds to five categories:

– its correlation to a primary frame, “material which already has meaning according to a scheme of interpretation, without which the keying would be devoid of meaning” [GOF 74];

– the fact that the alteration of the primary frame is known to the participants;

– the frame corresponds to the spatial and temporal bracketing described by Bateson. Goffman explains that indicators are made available for the beginning and end of these brackets;

– one can key any type of primary frame, natural or social;

– keying influences the definition given to the situation. This definition will differ between the primary and secondary frames: for example, in play-fighting, or between actual recruitment and recruitment simulation.

According to Goffman, events taking place in the two frames are just as real as each other (fighting–game), but in the example of the game, the action is in the non-literal secondary frame even while "it is literally carried out" [GOF 74].

Goffman identifies five basic keyed frames, according to which this transformation of the activities covered by the frame takes place:

– "Make-believe", very close to Caillois' mimicry. Make-believe is conspicuous for those who are witnesses or participants in it, and is relative to a less-transformed activity. According to Goffman, nothing is likely to come of it; imitation does not have the aim of being productive. He correlates "make-believe" in this sense to ludic activities and laughter, but also to fiction. Goffman underlines the importance of the nature of the frame with regard to the admissibility of what it conveys: "What is offensive in a movie might not be offensive in a novel" [GOF 74, p. 55]. He goes on to explain this admissibility, not in relation to the situation or the reference events (the "models") but rather in the type of keying itself. One could thus think that if "make-believe" in the sense of imitating hierarchical relationships, for example overauthoritarian or sexualized ones, might make colleagues laugh, this same imitation would be received very differently in a training role-playing exercise.

– "Contests" , which according to him regulates aggression and struggle. According to Goffman, the latter may be keyed by sport, but this will provide forms allowing distancing from the primary frame of combat.

– "Ceremonials": In *Frames Analysis*, ceremonials do not key life but an event, and Goffman insists that the people present at a ceremonial are not pretending to be anyone else, but occupy with a certain intensity their role:

ruling character, priest, spouse, etc. Ceremonials involve a dramatic dimension linked to representation, he notes here, as opposed to sport or fiction. It is clear, independent of the sacred dimension conveyed by ceremonials, what brings to mind the criterion of dramatization and the emphasis given to role – not here to the role played, but to the role occupied. In this double framing of play in work, the role occupied by a manager is not only a performed role but a real one. The social role of the person participating therein may be played, embodied as Goffman put it, in such a setting. Here, the role crystallizes a place that the person occupies in society, and which may be expressed here with force (much more so than in the solely play frame).

– "Technical redoings": This entails both imitations and simulations, but this time with no direct link to play or leisure. Technical redoings in fact have the purpose of simulating an event to learn about it or to experiment. Goffman thus refers to testing, training and repetition in this category. But also to the demonstration of know-how. According to him, technical redoings simplify or complicate the situations to which they refer in reality. Following the Goffmanian idea of an alignment of frames, games in organizations that are organized for training purposes will partly refer to this framework.

– Finally, "regroudings", which refer to the conduct of an activity for purposes other than the usual ones.

These keyings thus function according to "a set of conventions by which a given activity, one already meaningful in terms of some primary framework, is transformed into something patterned on this activity but seen by the participants to be something quite else " [GOF 74, pp. 43–44].

Simulation games rest upon this principle of keying the primary frame into the secondary frame. When children play knights and dragons, they refer to the model of a battle (primary frame), but make it into a secondary frame (the game – acting as if). Goffman gives an example of using a saw: when a person saws wood, this is a primary frame, but when the same person starts to saw on another person, it may be a primary frame of murder or a secondary frame of stage magic (in which the participants grasp, by convention, the artificiality).

One might consider that this is true for other forms of game than just simulation games, if we return to the categories listed by Caillois. For example, *Ilinx*: throwing yourself off a bridge is not the same thing as throwing yourself off a bridge attached to a bungee cord. There is a transformation of the primary framework, facing the void and death, into a secondary frame: making a game of it in complete (or relative) safety.

Goffman, pursuing his theory of frames also argues that one can "fabricate" frames, in the case of hoaxes or wrongdoing. In the theater, the fabrication of a secondary frame where, for example, someone in the hall begins to heckle the actors might sometimes raise doubts about what's happening: is it an actor, or has something from the primary frame broken through? (everyone is now uneasy).

But the intentionality is not only for creative or ludic ends. Goffman defines "fabrication" thus: "I refer to the intentional effort of one or more individuals to manage activity so that a party of one or more others will be induced to have a false belief about what it is that is going on. A nefarious design is involved, a plot or treacherous plan leading –when realized – to a falsification of some part of the world" [GOF 74, p.83].

Manipulating the frame, and in particular the context of a game, leads to the possibility of manipulating someone else. For Goffman, reality arises in some sense from a competition between meanings to give to situations. Con artists, jokers, forgers and fraudsters take advantage of this competition by using a frame coupled to a meaning that they hope will dominate the others. A show, like a scam, will emerge either as a success or a failure. Thus, for Goffman, "the act of perception" is "an integral part of the scene" [GOF 74].

This question of attempted mastery of the act of perception directly bears upon problems of communication, including managerial ones. Independent of the clear intentionality of manipulating others, the manipulation of frames offers certain benefits that interest, as we have seen, those who enact ludification as well as ludicization. In this latter case, transforming commercial goals and competition between colleagues to gain bonuses into a game (or "challenge") seems to bring us back to the question of Goffmanian "fabrication" of frames... just as much as in sociology, when workers themselves transform their activity into a game so as to endure it better, something we also see.

For Goffman, there may be "stratification" of frames, or their organization in layers. Thus he invites us to consider, among other interactions, a game according to this stratified reading: an individual asked to play a role-playing game in the company, are they playing, being trained or being evaluated? What will be the dominant perception of the situation, according to the place he occupies in the game?

1.4. The concept of play today

1.4.1. *The current syntheses of a definition of play*

Because he creates a very good synthesis after debating theories of play, we will end this survey of the theorization of play over the last century with that of Gilles Brougère.

Gilles Brougère is a professor of educational sciences and has worked on play, in a quite pioneering manner in France, for more than 20 years. The definition he suggests for play interests us in particular because it synthesizes the previous propositions, while refining them.

Gilles Brougère stresses first that there are two "traditional" ways to consider play and games:

– play and games as amusement or leisure;

– game as the form of an activity.

He thus distinguishes that play is produced affectively or psychically (amusement, leisure, etc.) from its formal characteristics.

One might stress, as he does elsewhere in his various works:

– that a formal characteristic of game does not yet make something ludic for those playing;

– that, moreover, playing a game is not necessarily leisure in the sense where, for example, it has been used for a long time in the framework of school for training children (and is today in the framework of training adults).

Play and games, he states emphatically, are a particularly complex object. We have already noted that the linguistic difference between *game* and *play* does not exist in French. We have also seen that Caillois attempts to reconstitute it with his categorization between ludus and paidia: the regulated and thus organized dimension of game (ludus) and the dimension of youth, disorganization, being out of control of play (paidia).

Game is all the more complex when it refers to various articulations: "A game can first of all be an object (and not necessarily a toy, he says) (...) a game can also be a set of rules and principles, some kind of immaterial object, like the game of chess (or playing tag). A game is finally (and doubtless most often) an activity linked to the fact of playing" [BRO 05, p. 7].

1.4.2. *Brougère's characteristics of play*

One last theoretician of play, important for grasping the theoretical framework of our research, is Gilles Brougère, who, in the tradition of Huizinga, Caillois and Henriot, refines the different articulations of play and games. His perspective is thus to concentrate on identifying general characteristics, rather than analyzing play from an interactionist perspective. As a prelude to his exploration of the concept, Gilles Brougère underlines the great complexity of play in terms of traditions of attachment to values which are sometimes conflicting. He thus refers to [BRI 97] and his reflections on the rhetorics around games:

– "the rhetoric of play as progress. This mainly concerns children's play. It defends the idea that animals and children, but not adults, learn to adapt and develop through play";

– "play as destiny, which applies to games of chance";

– "play as power, which relates to the domain of sport and competition";

– "play as identity, which refers to traditional games and ceremonies";

– "play as imagination, which applies as well to childhood creativity as to social activities of creation";

– “play as the self, in relation to solitary performances of pushing one’s limits”;

– “finally, play as frivolity, linked to old ideas of its association with madness, and repeated today in a critique of the modern vision of play” [BRO 05, pp. 34–35].

Without returning to the supposed operative force of play, perceptible at every level of these rhetorics, what Gilles Brougère suggests is a structural definition of play, linking game and play, independently of the diversity of the articulations to which it refers. One might think that he thus surpasses the attempts at typologies such as those elaborated by Roger Caillois, distributing the characteristics of play between *agôn*, *mimicry*, *alea* and *ilinx*. He thus addresses the conjunction of “conduct” and “situation”, as described by Henriot: “The key to the question of play thus relates to the establishment and articulation of the two concepts of conduct and situation. This double instrument must allow us to grasp the fleeting and ambiguous object that we seek to understand. For there to be a game, the situation must be ready for it. The subject finding itself in this situation must also have the ability to perceive and imagine the situation from this angle. Taken separately, neither the situation nor the mental attitude are enough for a game to be possible” [HEN 89, p. 216].

The characteristics of play refer Brougère to the attitude of the player (their psychic positioning with relation to the object) and the more structural elements of the game. According to Brougère, every game consists of:

– The “figurative element”: “A game thus appears as a second-degree activity” – says that play “is not for real” [BRO 05, p. 45]. Brougère cites Bateson and his reflections on the metacommunication inherent in games (communication about the current interaction), which he explains, in his own theory, as a figurative element. To make it known that a fight is in play requires the use and communication of a figurative element, which allows its own interpretation to be adopted and to resolve the ambiguity of the scene specific to play. “*A game is at the same time what it appears to be (...) and a game*” [BRO 05, p. 44].

The decision: “to play is to decide”. This affirmation greatly refines the question of freedom (freedom to enter the game) raised by the previous theories. This criterion profoundly nuances the nature of participation in the game and the space of freedom that constitutes it. Previously, the criterion of freedom was in fact thought of as necessary for a game: a game being thought of as, we might say, compulsorily voluntary. One might very well, especially in the case of a game within a work organization, but also in other contexts of existence, be put in a situation where it is difficult to refuse to play (coercion), and yet be ready to play. Brougère, introducing the criterion of decision, raises it at the point of entry to the game (to decide to enter the game might become deciding to play the game... or not, while still participating). Above all, he widens the criterion of decision by saying that "free entry into the game (...) is not however the most interesting aspect of decision within the game" [BRO 05, p. 51]. The criterion of decision according to him refers to the question of decisions made IN the game. In every game (and the theory of games is based on this principle), "individuals (the players) are led to make choices among a certain number of possible actions, in a frame defined in advance (the rules of the game), the result of these choices producing the result of the game, which is associated with a positive or negative gain for each participant" (Guerrien, *The Theory of Games, p. 5* [BRO 05, p. 51]). This leads Brougère to say that "playing is deciding"; deciding on the next action in a role-playing game, in chess strategy, in taking one's turn at cards, etc. The action of the game, bearer of this decision, takes place in a dialog or an adaptation, of some kind, with the decisions of the other players.

The game consists of rules: "To play is to decide to act in conformity with a rule, and it is at the same time deciding to accept this rule as support for my action" [BRO 05, p. 55]. These rules may be more formal (chess) or more flexible: as a child, playing mommies and daddies or teachers and pupils means agreeing to obey the behavioral codes of teachers and pupils, or of parental roles. Brougère stresses that rules may be altered or renegotiated by the players, and that this remains a game. The rules are worth not so much in themselves, but precisely because they have been agreed to by the players. This returns us, for example, to the reappropriations of games observed by ethnologists. Arjun Appadurai describes the acculturation of cricket in India (1996), but this has been observed, in an even stronger form, among the Trobriand Islanders [KIL 79]. Trobriand cricket, introduced by Christian missionaries hoping that the game would discourage the Trobrianders from their regular warlike practices, has radically changed the rules. For example, the home team always wins, there is no restriction on the number of players (from 11 in the original game, there may be 40 or 50 players per team), before the match the ball is blessed by a local religious leader who also asks that the weather stay fine, before dancing and singing that goes on quite long. The objective of these songs and dances is to promote the qualities and values of each team, while at the same time mocking them with sometimes sexual references. At the end of the match, there is dancing and feasting. What for a supporter of English cricket would no longer recognizably be a cricket game remains so for the Trobrianders. One might thus think, following Brougère, that the rules of the game are created *in situ*:

– Another criterion is what Brougère calls "frivolity", which he links to the absence of consequences: "the game (...) is constructed in such a way as to minimize consequences". "(...) while specifying: This does not mean that the game has no consequences" [BRO 05, p. 56]. In play – and we have seen it in animal play – it is possible to act "as if" without the consequences of the real reference situation: one may face an opponent in football or chess without killing each other, and if that happens, it immediately ceases to be a game. One may be an alien, some kind of animal or a god, one may participate in a murder party without anyone dying, etc. Yet, and this shows the finesse of Brougère's analysis, there may be consequences to the game, not inherent in its own frame but having an effect on other frames: one may neither lose nor win but become the subject, following the game, of derogatory or laudatory views. Goffman speaks elsewhere of evaluation games when the explicit as well as implicit goal is to detect such and such a weakness in the player. But one might more prosaically think of holding onto the bitterness linked to a defeat, or experiencing (for longer than the duration of the match) antipathy toward an opponent whose behavior was not appreciated during the game...

Finally, one last criterion of the definition of a game is uncertainty: Brougère links uncertainty to the fact "that the outcome of the game is unknown" (...) "Its story, even if provided with a pre-existing framework, develops as the game progresses." (...) "This is where the interest in a game lies, as opposed to ceremonies, rituals, or classical theater pieces" [BRO 05, p. 58]. This criterion is decisive in distinguishing play from ritual and is furthermore taken up by Roberte Hamayon in her anthropology of "playing": "(...) it is by the place allocated to this margin of realization that ritual is distinguished from play: everything is done in ritual to ignore it (except for making it, a posteriori, the cause of a failure), while everything is done in the game to exploit it. (...) The expected effect of a ritual is to do with its normative and rationalized character, which is also helped by its solemnity and sacredness. That expected of a game, in contrast, has to do with the randomness introduced to the context of the game by the progress of the game" [HAM 12, p. 317]. Uncertainty creates tension and suspense in the game. Thus, the interest of the player is maintained, even their concentration, which may explain in part the interest of the use of games in training. Participation and attention are also associated. This criterion of uncertainty seems equally intrinsically linked to the criterion of decision that, for the players, gives pace and direction to the game and guarantees the variability of the game, given the diversity of style and subjectivity of each player. This may appear strange in organizations: to give a role, not to chance, but to hazard, in its contemporary meaning of unforseeability and risk, is understood much better when carried out or thought of within the limits/frame of a game. It will therefore be important to analyze the margin allowed in this specific frame, and the relationship of this margin to the power of action of the actors, if we refer to the definition of power by Crozier and Friedberg [CRO 77].

Brougère disqualifies the criterion of pleasure, often invoked to characterize play, arguing that, on the one hand, many activities not relevant to play may be done with pleasure (including working!) and, on the other hand, one may experience displeasure in a game: finding the effort one makes to be taxing, conflict with other players, etc.

By characterizing play thus and nuancing or modulating the characteristics defined previously, Brougère defines play not as a concept with fixed limits but a concept allowing us to grasp an object in all its complexity and variability. He introduces the notion of degrees of play, allowing objects of hybrid forms to be described: some games are thus more games than others, and some criteria may not have the same force as others in a particular game. He thus speaks of "partial ludic characteristics", which makes play not a standardized but a modular object.

It is this complex, non-homogeneous object, linked to other frames, which we are to study as it takes place in work environments.

1.4.3. *The link with learning*

Gilles Brougère stresses the theoretical development, two centuries old, which argues a strong link, even thought of as "natural", between play and learning; play seen as being spontaneous to the child and referring, Brougère says, to the "myth of natural harmony" [BRO 97, p. 55]. He underlines the fact that "if play does not allow new learning, its contribution to development seems essential in that it helps entrench it" [BRO 05, p. 23], referring to Piaget. Play is thus thought of as contributing to the child's development, and creating the conditions for learning, as is implicit in pedagogical programs, particularly those of early childhood (kindergarten), since Froebel's original Kindergarten in 1836 [BRO 97, p. 50].

Play is also thought of as promoting informal learning, seen as unconscious or able to happen independently of the will of the child (or adult!) to learn: "Play is perceived as having a goal, functional implications which essentially escape awareness" [BRO 05, p. 29].

In this sense, play is operative or performative: it enacts actions of learning while not directly or explicitly having learning as a goal.

Brougère questions this, saying that this assumption, virtue of learning through play seen as proven (its proof very certainly the outcome of studies in ethology) has no systematic basis. In his work *Playing/Learning, he cites* studies that "show the impossibility of showing in a rigorous way any certain long-term benefit to play behaviors which by definition already appear devoid of immediate benefit" [BRO 05, p. 30].

Besides the fact that the potential of play for learning has no systematic basis, he stresses the differentiated relationship between play and learning: "It seems to me that the justification for play oscillates, without this being clearly stated, between visions of a vector for learning (it is through play that one learns), a context for learning (it is in play that one learns), and favorable conditions for it (it is around play that one learns, games allowing us to be available to learn)" [BRO 05, p. 75].

Finally, he insists, taking up deconstructionist theses concerning play, on the fact that children's play – though we might add to this the play of adults in the professional context – takes place "in a frame that in large part is determined by adults" [BRO 05, p. 77]; by which he means: those who prescribe games, parents as teachers. The type of toy given or the type of game taught to the child relates to the nature of the game – not free, but socially and culturally determined. A child plays different games, he says, according to whether they are a boy or a girl, according to their age, their environment, their country. And we can see that in businesses it is the same thing: different games are played, depending on the goals of the context in which the game occurs, whether played by the managers or the workers, the engineers or the salespeople.

Brougère returns to the fact that play is thought of as natural even though it is an artifact, a construction: "(...) what is thought of as natural is in fact a cultural artifact, what sees itself as universal is local, what sees itself as freedom is control" [BRO 05, p. 84]. His comparative work on preschool systems, and more generally his research on play, show how ultimately play is conditioned according to what, as a bearer of cultural, social and professional norms, one thinks play is. People according to their cultural inscription or their environment are carriers of an image of what a game is: for some, a "challenge" in a company will be a game, and for others it will mean having to set possibly tightened objectives for work, productivity or performance. For some, role-playing games around their activity will be fun; for others, they will be something of a humiliation.

We must then, in the professional and managerial context that we intend to study, perceive norms, categories and frames in/according to which games are carried out. We must also address play according to the variable nature of its characteristics, as analyzed by Brougère. The choice of analysis that we enact will aim to grasp it according to two structural traits:

– its dimension of "margin", a game-space between several frames, spheres of interpretation that will lead us to study the relations between them, as well as those elements pertaining to these frames and interpretations;

– its performative scope, interrogating what a game does in taking place.

But before that, after this survey that we have carried out of theories of play, what is left of the traditional (and factitious) opposition between play and work? What games are mobilized in the work organizations which we aim to study? And finally, are they games?

2

Games in Business

2.1. Relations between games and work: an apparent incongruity

In the theoretical literature about games, differences between games and work, and the existence of times and spaces dedicated exclusively to one or the other have often been pointed out. These two activities are regarded as contradictory to the extent that a game, which would become work for a professional player, would no longer be considered as such [CAI 67]. The first activity has to do with leisure time and the second is incompatible with games: "[a game] always results in an atmosphere of rest and entertainment. It relaxes and amuses. It recalls an activity without any commitments and without consequences in real life. It is opposed to the grave side of the latter and therefore it is accused of being frivolous. It is also opposed to work; lost time against utilized time. A game produces nothing: neither goods, nor products. It is essentially sterile" [CAI 67, p. 9].

Among the features listed mainly by Huizinga and Caillois, and then regularly mentioned again, several are inherently opposed to the concept of work: the "freely agreed rules" or the freedom to start and finish a game, its gratuitous nature and lack of productivity, or the fact that it is "autotelic" and has no other goal but its own enactment [BAL 02]. Naturally, its components of "delight and enthusiasm" or "the joy and relaxation" that characterize it, according to these authors, represent yet another reason to consider *a priori* the two activities as diametrical opposites.

Roberte Hamayon revisits in a footnote this opposition, which is "widely acknowledged by Western authors basing their work on games" [HAM 12, p. 46], and consequently is not relevant in other parts of the world: "this association stumbles upon the evidence of linguistic associations between work and ritual elsewhere in the world" [HAM 12]. Hamayon underlines the possibility of interpreting games and work as sharing several common traits with ritual, seen as an anthropological concept.

However, other characteristics such as the individual and social dimensions of a game, the uncertainty it entails, as well as the tension associated with achieving success or winning, could at first challenge this traditional opposition between the two terms.

As we have seen, more recent analyses of games (Brougère) are improving these "historical" definitions: a game is associated with a figurative dimension, it generates a decision-making process (for the player in the game) according to mechanisms that govern the decision (rules), it is half uncertain and half frivolous and finally its actions would not lead to the same results in a non-game context. Most importantly, these elements should be interpreted in relation to the marked unpredictability that characterizes their force and presence in the game. This new classification of the features of games results in changes in the related terminology that attenuate the diametrical opposition between games and work. This redefinition prevents the paradigms of games and work from being set, in a way, by their mutual opposition. Brougère also emphasizes the marked similarity between children's games and adult work, and he regards games as an introduction to work: "It is indicative that games turn into a school exercise or work, and not into adult games or leisure time. Children's games seem more similar to adult work than they are to their adult counterparts" [BRO 05, p. 24].

Henriot had already pointed out that games and work are close activities:

> "If we consider the psychological dimension of the term, there is work in every game: attention work, perception work, muscular work, intellectual work – together with those elements that any kind of work usually involves: effort and strain. In all likelihood, this is the reason why, in terms of behavior, it is difficult, and even impossible in certain cases, to distinguish between games and work" [HEN 89, p. 197].

Consequently, the conflictual nature of these two activities must be reconsidered. Several authors, especially those writing about the sociology of work, have been undermining this opposition for several decades with studies conducted in the workplace. Burawoy's works [BUR 79], carried out in a mechanical engineering plant, Sherman's studies [RAC 07] in the hotel business, and the works of Frenchwoman Marie-Anne Dujarier [DUJ 15] about senior management show the connections between games and work. Games are actually used to do the work. They allow playful relationships on the job as well as fighting against boredom, involvement and disengagement, sometimes unawareness of the consequences of one's work on what lies outside the game context, like the impact of one's work on employees or customers. Turning work into a game may consist of creating "a collective construction of a certain 'reality' which helps us manage to achieve it" [DUJ 15, p. 218].

Michael Burawoy is an authority on the relationships between games and work. Taking over from another sociologist, Ronald Roy, after 30 years, he studied changes in the "regimes of production" of a metalworks plant. According to him, these regimes of production have historically developed from the burdensome constraint on the workers' consent. Consent and obtaining consent are particularly significant notions when we study present-day work and management practices and we can see in Burawoy's work the beginnings of the studies about the "fun work environment" and "gamification". The aforementioned blurring of the distinction between work and free time and the creation of a continuum in businesses in order to organize work as a game contribute to this consent. Burawoy mentions "compensations" and "relative satisfactions" which, according to him, constitute the framework of games. By "relative satisfactions", he means, besides contentment, something which may be closely related to a form of (self)-domestication, namely tractableness and inurement [BUR 79, p. 78][1]. Becoming tractable and accustomed represents a completely relative pleasure that is, however, essential to the fact itself of being able to work, which is the goal of both workers and directors. Burawoy, unlike several of his predecessors, underlines that games are not organized against management.

1 "Work realities (physical conditions, repetitiveness and routines) give rise to deprivations (impairment, tedium and weariness), and deprivations engender relative satisfactions (inurement, traction or tractableness and contentment). [BUR 79, p. 78].

> "Rather, they emerge historically out of struggle and bargaining, but they are played within limits defined by minimum wages and acceptable profit magins. Management, at least at the lower levels, actively participates not only in the organization of the game but in the enforcement of its rules. The stimulus to engage in such work games derives as much from the inexorable coercion of coming to work, and subordination to the dictates of the labor process once there, as from the emergence of "radical needs", "a new vision of work" (…). The game is entered into for its relative satisfactions or what Herbert Marcuse calls repressive satisfactions. The game represents a need that is strictly the product of a society "whose dominant interests demand repression". The satisfaction of that need reproduces not only "voluntary servitude" (consent) but also greater material wealth [BUR 79, pp. 80–81].

According to this author, what a game involves is "a set of rules, a set of possible outcomes, and a set of outcome preferences" [BUR 85, p. 38]. Games seem to have a transforming power that reduces constraints and some of the demanding or boring aspects associated with work. They seem to represent here a repressive measure against the reactions of refusal or rebellion against the task that has to be carried out, while also allowing employees to get involved in their work again. Games are used to make this possible or bearable. Far from being opposed to work, they actually complement it: they represent a solution adopted so that work can be done. Consequently, Burawoy points out that not only middle management contributes to them, but that directors also let them take place. There is a sort of symbolic compensation related to games in the workplace, which reduces boredom, helps adherence to the work rules conveyed by the game (or made compatible with it) and keeps workers attentive. Burawoy, taking up Roy's works and leaving behind the notion of "relative satisfaction" (which he uses from a critical standpoint, even if we may be led to think, as far as the operators observed are concerned, that it is possible to get caught up in this), mentions accomplishment, self-expression, victories, stratagems, applied knowledge and skills. Sherman will say that games favor "skill, control and autonomy" [SHE 07, p. 151]. We can deduce from that why management will tolerate their use.

According to him, who in turn follows and quotes Jason Ditton, some characteristics of games, such as the creation of uncertainty, strengthen management power. Management have a vested interested in having workers regain control over machines instead of letting themselves be governed by them in relation to the autonomy they acquire. What Burawoy notices appears to be the worrisome beginning of the new management beliefs (we are considering the industrial environment of the 1970s and not the leisure industry of 1990–2000) that advocate the autonomy of self-managing employees aiming for more involvment and flexibility. Management has not started instructing employees to take the initiative yet, especially in the case of workers in machine shops. What actually takes place, through the dynamics of voluntary submission enabled by the game, is that employees take control, which favors productivity. Another benefit of games pointed out by Burawoy is the deflection of conflicts, given that immediate supervisors agree to their practice. A colleague becomes the adversary, the person we should fight against or compete with: "[...] the combination of autonomy with respect to machines and dependence with respect to auxiliary personnel has the consequence of redistributing conflict or competition" [BUR 79, p. 81]. In this sense, Burawoy refers to games as something that brings harmony to the workplace (...as they are not reflecting a preexisting situation of harmony).

2.1.1. *A variety of ways to address the relations between games and work in the social and human sciences*

The relationships between games and work can be interpreted in different ways, if we actually consider what has been written in the social sciences about them:

– The way workers regard their activity as a game:

As we have seen, this may allow employees to find a meaning in their work or to cope and deal with it. Games may also – even in this case they are tolerated by management – allow employees to remain as attentive as they need to in order to work [DEJ 93, DES 91]. Dejours and Dessors describe the games they observe in a petrochemical plant where supervisors in the control room regularly play Scrabble. While they note that the players are ashamed of their game and hide it, and that management is aware and disapproves of it without, however, sanctioning it, they underline the significance of the game which creates, by preventing boredom or distress, favorable listening conditions for the good maintenance of the plant: "This is how the workers have conceived a "secret", a "trick" to control the process effectively. Using the body to examine the process is difficult. If workers start actively listening and thinking about it, while focusing on the noise, they can no longer make out any sound. Either they cannot hear anything anymore, or all noises become suspicious, they become confused, and they soon become anxious. They can no longer use their senses. The cruising production speed in a way requires workers to relax and operate on automatic pilot themselves. Then, they can get physically and sensorially in tune with the system and identify without hesitation the anomalies that take place during their watch. In this context, we can see in hindsight that playing Scrabble is "genius"! Strangely, they play Scrabble instead of belote, which is a far more common game among French workers. When people play belote, they actually talk a lot and make noises. During a game of Scrabble, everyone is quiet. While getting rid of boredom and distress, the game of Scrabble enhances sensory performances. The game reconciles a quest for comfort with technical effectiveness". (Christophe Dejours, "Practical intelligence and wisdom: two unknown aspects of actual work", [CHR 93]). Once again, we find the idea of synergy between a game and work as established by workers. This leads us to directly analyze the type of game used in a specific context. We can identify in the remarks of Dejours and Dessors the correspondence between work and play activities established by choosing Scrabble rather than belote. The same can be said about the workshop studied by Burawoy, where games are related to production rates and quotas. In Dujarier's reports about the planners' work, the issue consists of facing a challenge as well as finding a strategy to fire 500 people without provoking social unrest. Rachel Sherman, while studying the luxury hotels business, underlines how the games played in the environment she studies vary from those used in the industrial contexts observed by Burawoy: "Games take place in a collective context, and they are linked to workers' status relative to that of other workers. Workers require some autonomy in order to play games, so highly controlled and routinized workers doing repetitive jobs are unlikely to participate in the kind of games I am describing (Housekeepers, for example, employ different strategies in their work but they do not play games with variable outcomes)" [SHE 07, p. 111].

Moreover, these games vary in relation to the hotel considered. Each type of job has its own game, which corresponds to Brougère's remark about how the games organized by adults for children during the learning process or for their alone time are socially and culturally defined. Regardless of any immediate goals, which – as we have seen – can be questioned [CAI 67], is learning to play consequently related to learning to work? In this case, which elements that can teach us how to work, as an activity, could games involve? It will be interesting for our research to analyze from this perspective the use of games as introduced by management and not organized by workers.

– In organizational theory, the actor-player's notion of strategic action (Crozier/Friedberg):

Games are interpreted by these two experts in organizations as strategic actions that are part of a power struggle. Power relationships are governed by the strategic role played by the actors. They draw their inspiration from game theory and the very famous prisoner's dilemma to conceive the actors' games in a work organization. Crozier and Friedberg regard power as an edge that any actor in a business can have – in varying degrees – or lack: "Power lies in the margin of freedom available to each partner in a power relationship, i.e. whether or not this partner is in the position to refuse what the other one asked" [CRO 77, pp. 69–70]. Seizing power consequently amounts to: "altering the nature of the game, or changing what is at stake and shifting the uncertainty zones, making the most of the circumstances to push the other towards a much less favorable territory or lead him to give up" [CRO 77, p. 71]. The game corresponds in this case to the creation of uncertainty, which modifies the power struggle. It is a piece of information that gives way to a decision, part of a relationship where the aim consists of widening one's margin of freedom and reducing that of a partner/adversary". To use their expression: "the strategy of each partner/adversary will naturally consist of manipulating the predictability of his own behavior as well as the other's, either directly or indirectly by modifying in his favor the structural conditions and 'rules' that govern his interactions with the other. In other terms, he will have to increase his own margin of freedom and arbitrariness as much as possible, so as to enjoy the widest range of potential behaviors while also trying to restrict that of his partner/adversary and restrain him with such constraints that his behavior will, on the contrary, become perfectly predictable" [CRO 77, p. 72]. In this case, organizations and games are read in an agonistic way and, for example, themes of cooperation or community are left aside. Contrary to the issue of predictability in the eyes of the other, linked to Goffman's "facework", in this case what matters, in order to take the upper hand in a certain field, is unpredictability and its related issues. While Goffman focused on the effectiveness of codes and metacommunication in relation to experience so as to be able to function in society, here the issue concerns changes in the rules and garbled signals. In organizations, individuals play a part (metaphorically represented by a game) in opposition to someone else, where the quest or main gain consists of taking control of the situation. Crozier and Friedberg put forward a "dualistic" interpretation inspired by the concept of game. According to them, an organization has to do simultaneously with "the self-centered strategy of an actor" and "the finalized coherence of the system" represented by the organization, experienced here as the "result of the game" [CRO 77, p. 237]. These two aspects, namely the actors' strategies and the coherence of the system, may potentially structure work organizations and cannot be considered independently from each other. However, a reading that focuses very specifically on the players-actors may conceal more complex realities.

– "Playbour" [KÜC 05]: Kücklich starts from an analysis of "modding", which designates a practice where videogame players develop new ways of using a game (modifications, additions, etc.) free of charge, ultimately benefitting the software publishers. "Playbour", the use of gaming practices, gets to encompass other ways of exploiting the players. Thus, the term refers to so-called "farming" techniques (the collection of gold and virtual points to make money) employed in "virtual sweatshops [DIB 07, GOG 11]. In this case, gaming practices are industrialized for particularly restrictive and repetitive production purposes and very low wages. This follows on from what could already be noted in relation to the work of testers in the video game business. These are actually poorly paid jobs that require mostly employees with low qualifications – who are, however, very skilled in terms of their intensive use of videogames – to replay unremittingly the same game sequence in order to ferret out "bugs" (when this activity is not assigned to players themselves in case a game is launched on the market without being actually finalized in order to meet the deadlines set by the sales department).

– The management's organization of work as a game:

On the one hand, we have to consider what we have called "laicization", i.e. usual work activities presented as games and the challenges faced by sales executives in order to replace the more traditional concept of "sales quotas". On the other hand, the management's introduction of a game element leads us to the goal of this quest. It is called "gamification" and designates, as we have seen, the use of game structures in a context foreign to games. We should point out that the last two aspects that research on the relationships between games and work focused on (playbour and gamification) are the most recent. To use Mollick and Rothbard's expression [MOL 14], ludicization and gamification follow the logic of "mandatory fun". However, unlike "work design" (managing the work activity and/or workstation) or "job crafting" experiments (changes made by workers to their job or activity)[2], gamification has to do with work experience rather than the nature of the task, as these authors point out. We can add that the difference is that if management aim for work experience when they gamify work, what has actually changed is not necessarily the experience of the employee when he becomes a player. On the other hand, the structure of the task is objectively improved by the game, as we will show in this study.

We will not consider here those games played by certain employees in the workplace that lack any aspect of cooperation or relation to the activity itself. These practices are in line with the opposition we have mentioned. Play time differs from work time and the latter does not tolerate or organize it. In other words, the relationship of exclusion between the one and the other holds true.

2.2. The game in business: returning to a typology

If we consider the four-category classification (agon, mimicry, alea and ilinx) established by Roger Caillois, it is competition and simulation games that are the most common in a business environment[3]. We notice that gambling seems to be excluded from it, even if it is possible to draw lots to form teams, in order to determine who will start the game, or cast dice to reach a new "square" in a board game. Games that induce lightheadedness and dizziness are limited to group sessions of bungee jumping, introduction to gliding, etc., organized during seminars and far from the office and the everyday life of organizations. On the other hand, competitions between individuals or teams and the use of simulation, especially during training or team building processes, are frequent. As for the former, all kinds of games

2 Both aim for improved job satisfaction.

3 This observation has been made on the basis of a web review of the game services offered by communications and consultancy firms.

may be employed: sports competition, card tournaments, competitions between teams building Lego towers, etc. The latter, on the other hand, focus more on the notion of role and the simulation of “scenarios”. Finally, there may be hybrid forms. For example, serious games (SG) can use challenges and scores, by means of avatars, in relation to the character or function considered. As a result, games belonging to the two categories of agon and mimicry can be:

– “Challenges”, or competitions between teams or employees set up to achieve goals that are generally assessed (in terms of volume or time).

– Simulation games that can take different shapes, as we have been able to see: “role-playing” during training courses, “reversal days” (or “Try my job”), business theater. In our study, we focus on those games that are designed to involve work: management task training, interaction work, implementation of intercomprehension tools, cooperative solution finding, etc. We will consequently leave out role or simulation games that do not directly involve work. We can point out that in this case the difference between “play” and “game” helps us frame our topic. Let us consider a “murder party” when its goal is to build a team involving an element of shared enjoyment (fun) by using a game (play), or when the structure of the murder party game is used to formalize explicit professional expectations of the participants (game/gamification).

– “Serious games”, which will combine these forms and create, unlike the other three categories, a man–machine interaction that does not necessarily allow the intervention of a third party, as well as implement a system that most often personalizes the “game” (cooperative SG are not the most common).

– The introduction of board games, cards, Kapla, Lego as well as the game of Go into work contexts, so as to employ them directly (for example converting a board game like Trivial Pursuit into a question–answer type of game about skills) or indirectly as tools during practice sessions or professional training, through the figurative reading or analysis of game behaviors. The competition and simulation results can also be combined in this case.

Here, we will not present any classification of the games used in work organizations. We will use instead the broad categories of games put forward by Roger Caillois. Employing anything more specific than general categories determined by broad game principles, such as those proposed by Caillois, turns out to be difficult due to the large number of games used in business environments. Millick and Werbach [MOL 14, p. 439] underline how "the range of functions to which games are applied and the variety of gameful approaches used by enterprises make it virtually impossible to develop a complete synthesis of enterprise games in a theoretical sense". This will lead them to classify games in relation to the kind of results expected by management[4].

Consequently, we will first mention the different types of games we came across in our research in order to illustrate the broad categories of agon and mimicry, which include competition and simulation games. Second, we will focus on simulation and role games, namely on the specific category of mimicry, which will back up our analysis on how management uses games. We should point out that such games, in particular role and simulation games, are favored to train and instruct managers just as challenges are chosen for sales executives. Management schools regularly use management simulation games (business cases and business games) for their training, as Léo Touzet [TOU 13] underlines, which leads us to think that resorting to them is part of a standard and well-known training method for these groups. Besides, we will focus on "indoor games", i.e. games organized within organizations and during working hours, namely those games that form an integral part of work[5].

2.2.1. *Challenges*

A "challenge" in a business can be of different kinds according to its goals. An agon encompasses both team-based sports or cultural challenges in teambuilding processes that aim to reinforce interemployee relationships and a commercial competition that combines a logic of goals with material or symbolic gratifications. These gratifications may consist of bonuses, gifts, or more symbolic benefits: being nominated the best employee of the month or, like in the Carglass business, endorsing the recording of the company's radio commercial. As for managers, challenges may have to do with finding better strategies or solutions. The challenge dimension can be regarded as, at least,

4 This topic will be dealt with at the end of this work, which is dedicated to the functions of games.

5 Unlike "outdoor games", which are generally more recreational (treasure hunts, paper chases, sports competitions, etc.).

double: as with certain competitions based on "hackatons"[6], a challenge consists of both creating competitions between teams or players and being able to hold out for long periods of time during the competition and confront oneself. Performance is the key word and it consists of three aspects: "This is in essence the overcoming of obstacles: the pursuit of a goal, the rewarding of progress and the presence of compelling challenges that demand sophisticated strategies" [CHA 10, p 50].

The relationship between businesses and sports competitions seems to date back to the beginning of the 20th Century, according to researcher Béatrice Barbusse: "In the 1920s, we see the appearance of corporate tournaments that opposed trade unions like postmen, railroaders, metal workers,..." [BAR 02, p. 402]. From a corporate activity, sports would later become "business sports" [BAR 02]. However, Béatrice Barbusse mentions that the relationships between businesses and sports involve much more than mere competition. They consist of ensuring the employees' fitness and healthy lifestyle, as well as inviting trainers and coach to the firm to give speeches about their management experience. They also include the use of sponsoring techniques, in order to advertise the organization, and the repatriation of sports celebrities. This researcher puts forwards a classification of sports-related activities in a work context:

Sports in business: several different uses

Sports competitions

– Inter- and intrabusiness competitions: business sports (formerly corporate sports), extra-corporate competitions (Business Cup, *Challenge du Lys*, Challenger's Trophy, CEO Creathlon, business *Cross des Violettes*, business *Cross du Figaro*, business Olympics, *Défi charantais*, interbusiness rafting Open, *Trophée du Dauphin*, etc.).

– *Grandes Écoles*/business competitions (Mont-Blanc Challenge, Eurochallenge, the Olympub Games, the EDHEC Sailing Cup, Montathlon, *Spie Dauphine*, *Trophée des Battants*, etc.).

6 At first, they were collaborative marathons involving software programmers. They have inspired, for example, the HRackaton (www.hrackathon.com, March 2015), where teams of students, HR staff, and IT engineers (Axa, Danone, Société générale) are made to compete for 48 h in order to develop a recruitment app.

Seminars, work experiences:

– motivational (extreme internships, outdoor work experience);

– management training.

Fitness rooms:

– fitness spaces, cardio-fitness centers, etc.

Sports metaphors:

– verbal (sports rhetorics);

– visual (sponsoring).

Recruitment:

– top-notch athletes;

– *"sports employees".*

[BAR 02, p. 405].

In all the scenarios considered, Barbusse underlines that "each time, the goal does not change much. It is a matter of giving a clear and precise meaning to the way a business works, namely to indicate, on the one hand, the way work must be organized and, on the other hand, to define everyone's place and role" [BAR 02, pp. 404–405]. Sports convey standards such as excellence, pushing one's limits, quickness, cooperation in a team and competition. It also translates certain organization principles: from a boss to a trainer, including players and the role played by each within the team.

Béatrice Barbusse, on the basis of the analysis of the managements' speeches about sports, mentions that sports involve such "qualities" as "a good physical condition" and that they are considered a positive way of "shaping one's character" [BAR 02, p. 407]. Sports develop individual qualities like toughening up, a liking for challenges, cooperation and performance: "they strengthen or develop a fighting spirit, drive, courage, competition, loyalty, responsiveness, sociability, and responsibility; they encourage the creation of personal mental steadiness (self-confidence, self-control, assertiveness, etc.) and facilitate the formation of team spirit" [BAR 02].

"Challenges" such as those set up in companies, in agreement with the spirit of sporting competition, follow the logic of individual as well as collective performance: productivity, speed, and (apparent) compliance with rules are most often integrated into the assessment of the winners of the "challenges" proposed to the employees working for these organizations.

Challenges are quite generally of a commercial nature: competitions between sellers, in which the person who has sold the most (winning points, bonuses, setting the winner up as an example in the workplace, etc.) will be the winner, or service employees, such as those dealing with customers, so that the quickest and most efficient in processing requests will be hailed as winners. They also aim to favor team-building processes by making sure that competitors, arranged in teams, meet and cooperate to make their own team win. They can also be used to create competitions during recruiting campaigns, such as the "E-strat challenge" set up by L'Oréal, which makes teams formed by three students compete on a global scale in an online strategy game for 7 weeks. Recruiters are met at the end of the challenge, when the finalists are invited to Paris.

Challenges introduce two elements: evaluation – generally by assigning points/scores/levels/progress bars – and gratification. In this context, gamification has been criticized and regarded as a mere "pointification" [ROB 10] or corresponding to what Félix Raczkowski [RAC 13] calls "scoring economies", which is a perspective that reduces the concept of "game" itself and can lead, according to this researcher, to the behaviorist practices used in 1950s psychiatry related to the use of tokens. However, if we leave aside studies on games, and focus in the strict sense of the term on research about gamification, assigning points, creating competition, and the evaluation indirectly related to the process have, on the contrary, been widely quite visibly improved:

> "[G]ame technologies excel at nothing so much as scoring, comparing and rewarding progress [...]" [CHA 10, p. 199]. Jane McGonigal considers scores to guarantee the productivity enabled by the game: "The more points you earn, the higher your level, and the higher your level, the more challenging work you unlock. This process is called "leveling up." The

> more challenging the work, the more motivated you are to do it, and the more points you earn...It's a virtuous circle of productivity" [MCG 11, pp. 1123–1125].

The critical importance of assessment and evaluation of the performance given does not seem to have the same impact or remain the same, according to whether we consider challenges, SG, and even board games, on the one hand, and simulation games, on the other hand. However, we will see that evaluation plays an equally crucial role for the latter: adopting the right or wrong behavior, managing to change the initial situation, overcoming the destabilization brought about by a new role or the new situations – even the imaginary ones – faced, etc.

2.2.2. *Simulation games*

Simulation games, which belong to the category of mimicry put forward by Caillois, have to do with the "not for real" aspect of the games that ethologists focus on. They combine learning and action, and their function can be interpreted as the reduction of this dichotomy to these two terms: learning places and times, and action places and times. Crookall and Thorngate [CRO 09] suggest that this reduction is thought to work in two different ways: it involves the use of one's knowledge to become more efficient and the increase in one's knowledge through action.

Simulation games in the workplace – and undoubtedly in broader learning contexts – aim to allow players to become reflective. Reflection highlights the educational need to think about one's action after it has been performed as well as when it is being performed. This pressing need further underlines the fundamental idea of an action that is disconnected from the reflection about it, so that this disconnection justifies the use of tools that lead to reflection. Crookall and Thorngate [CRO 09, p. 22] underline the cultural dimension of this dichotomy, which is specific to Western ideas about knowledge: "In many aspects of our everyday life, we hardly distinguish between action and knowledge; we proceed as if they were one. However, Western education has distorted life".

In a professional environment, what is an action without reflection? What does that tell us about how workers are regarded? Or, more precisely and hypothetically, what does that tell us about the expected (and no longer "wild") reflection about actions, in specific times and with particular tools, which could be governed by simulation games? Work psychodynamics actually define work by including subjectivity and the subject's reflection about his action, which he is led to constantly reassess and adjust so that the task he has been assigned can be carried out. This interdisciplinary approach to work then provides this definition: "Human work relies on the use of expertise and original skills, since it is required exactly when conventional knowledge and technique turn out to be insufficient to ensure the mastery of the work process" [DEJ 12]. This definition follows on from Davezies' – whom these authors refer to – which regarded work as what an individual at work does to overcome situations that have not been foreseen by the instructions he was given. Thus, they put forward the notion of a work process that cannot take place without the use of our reflective abilities.

Simulation games not only allow us to link reflection to action, but also isolate the latter from the consequences it would have in the actual world. Crokall and Thorngate [CRO 09, p. 22] define it as "inexpensive", which means that it involves a limited number of risks. Being able to carry out experiments and implement new forms of learning in the simulation space-time would limit risks, which may implicitly underline the negligible part played by creativity or testing in the real world. Testing and creativity should be "isolated" in a virtual dimension. The actual world would only be characterized by proven safe-mode actions, in contrast with the principle of variability well known in ergonomics. This principle enounces that for each new operation, we add new parameters, events and variations, which require us to have recourse to the workers' subjectivity and inventiveness in order to cope with mishaps or variables at work.

2.2.3. *"Real-life scenarios" or "role-playing games" during training*

Although "challenges" may be made up of (such as the E-strat challenge conceived by L'Oréal) as well as integrated into the actual sphere of work activities (sales competition), real-life scenarios are a fictional tool by definition and belong to the category of simulation games, and more precisely to the class of role-playing games we have observed.

Some researchers (like [FEI 02]) suggest that we clarify the forms of the simulation games used in organizations and targeted to adults by underlining the broadness – and consequently, in part, the vagueness – of this category.

If a simulation follows a model (we can simulate a context, certain behaviors, interactions, etc.), it can take, according to them, several shapes: verbal, graphic, mathematical, and many more. Simulation may involve individuals, computers and visuals.

Real-life scenarios, also called simulations by trainers, represent a so-called "immersive" technique in which participants are invited to "act out the role of a character or part in a particular situation. The participant follows a set of rules that defines the situation and then interacts then with others who are also role "playing participants" [FEI 02, p. 59].

Role-playing is a social game, unlike other simulation games (flight simulators, simulations in SG, etc.). According to these authors and other specialists, it is this aspect that gives us "an in-depth understanding of many of the social interactions that arise when evaluating or solving a problem" [FEI 02, p. 59]. If the roles and the scene played are fictional[7] the interaction and the events that unfold are real. As we will detail later on, we face the problem of finding out to what extent the "not for real" aspect of mimicry and the "for real" dimension of work interactions are similar. The developers of these games or the trainers that employ them often specify that role-playing confronts the interacting individuals with a situation "close to" what they may encounter in the workplace. This notion of closeness and game highlights the contrived nature of this process aiming to recreate a predetermined situation which is, on one hand, "standard", on the other hand, necessary for learning and development, and finally involving what professionals come across in their daily life. Role-playing as a structure and hypothesis refers to the choices made to enact it, which are suggestive of the definitions assigned to the work situation and those involved in it.

The specific form of simulation at the center of role-playing involves participants playing the game and embodying a role. Unlike preprogrammed software simulations (where computers suggest a series of steps with limited choice), role-playing involves a lot of improvisation. However, Feinstein

7 Like Gilles Brougère, let us recall that although some games may be fictional, none of them are imaginary.

et al. identify two of its limitations: the "feedback" of the interacting parties involved in role-playing may be different from their real-life equivalent (individuals play, provoke, and react in a way they would not in their actual workplace) and participants, because of how they are chosen for the game and due to the variety of departments they work in, etc., may not be equally or even aptly skilled when facing the situations envisaged. They may be "poorly equipped to respond in a manner that is congruent with the objectives of the learning activity" [FEI 02, p. 60].

Feinstein *et al.* emphasize how simulation tools are used much more for training staff than they are employed for "educational" purposes. In light of the training sessions involving "real-life scenarios" that we have described, our opinions about the distinction between training and learning are much less clear-cut. If role-playing is employed to train professionals for the situations they encounter, it is constantly preceded or followed by didactic briefing (for example on psychosocial risks, bringing employees into line, the relationship with patients and with their family, discrimination, etc.). Followed by "debriefing" sessions during which all the participants are invited to make comments on the role-playing activity, they make people think, lead them to change their actions, like in a training session, as well as make use of the theoretical elements provided during training.

The common development of role-playing games is divided into:

– A preparation phase during which the players receive written instructions for each role, which they do not share (a player knows nothing about the instructions given to the other players).

– Role-playing itself when players are introduced in the order that has been assigned to them. The game comes to an end if the players decide to stop it or when the organizer/trainer decides to end it.

– "Debriefing" during which those observing the game, the players and the organizer will comment on it by considering the moments they have deemed the most significant, in terms of the goals, set as well as the behaviors adopted or the events driven by the players. The trainer encourages certain reactions, promotes counterreactions and eventually intervenes by making comments. Debriefing, as Crookall and Thorngate [CRO 09] point out, is based on the principle that knowledge can result from an action– role-playing is commented upon so as to learn something from it. Preceded by a teaching session, role-playing also assumes the action to result from the knowledge acquired. There is then a mutual relationship between knowledge and action. Debriefing is supposed to focus on the moment of reflection and also to establish a connection between the action performed and the contribution of knowledge behind it. However, as these two authors point out, the three phases of teaching, action and debriefing are distorted or transformed: "We teach knowledge K, measure action A, and find that A does not reflect K" [CRO 09, p. 18].

Simulation reveals two gaps: the one between reality and simulation as well as the one between knowledge and action. We will see how debriefing, and more generally the effect of the comments made by the observers and the organizer, tend to bridge these gaps and attempt to bring about some regulation.

2.2.4. *"Reversal days" or "Try my job"*

Several firms and administrations propose, within their departments, day-long job and role rotations between bosses and employees or work colleagues. These events take place in French organizations and are called "*Vis mon job*" (Try my job) or "*Vis ma vie*" (Try my life), which refers directly to a mainstream TV program in which a boss is followed by cameras while he tries out different workplaces in his own firm (generally menial jobs). This process is also called "reversal day" or "role reversal day", and finally "job rotation". In France, we have listed three main processes called "reversal day" by organizations: work groups where a professional explains what his job consists of to one or more of his colleagues without any actual job rotation; job reversals between employees and bosses for a day with ad hoc work arrangements (extra number of people, working in partnership, etc.), and finally an actual job rotation between colleagues in charge of different tasks and including managers (only one case found). In this work, we are interested in the last two processes, since they follow the logic of role-playing.

The kind of job rotation employed by Anglo-Saxon management designates two main kinds of workplace swaps: the first one aims to limit the tiredness or boredom associated with certain tasks [CAM 94], without forgetting its use to fight against the onset of repetitive strain injury in the industry [JOR 05]; the second is designed to train and teach about the "different stages taking place when a product is created or a service is provided" [HUA 99], but also to provide retraining schemes, career development, and "employability" [CAM 94]. The timescales for each of these operations are very different: from a work-day rotation to make things less boring or standing in for an absent colleague, to 6 months to 2 years in a training context [CAM 94]. Job rotation has been met with enthusiasm by North American managers since allegedly 24% of organizations employing more than 50 individuals were making use of it at the end of the 1990s [GIT 98]. Intrinsically, this method has nothing to do with games and, for that matter, it is not presented as such. It matches well with the kind of organization of work that simulaneously prioritizes risk prevention, the identification of everyone's role in a production process, and flexibility, even if the concept of "job satisfaction" is not foreign to it [HUA 99]. In France, a "reversal day" (following the example of the Anglo-Saxon "role reversal day") combines the dimensions of training and understanding the work and role of others with the expected mutual understanding within a team, which coincides with the concept of "job rotation" but also includes a process of hierarchy reversal, directly inspired by reality TV and more carnivalesque forms[8]. It is most often initiated by internal communications services or human resource departments and it is proposed by certain advertising agencies as an internal communications event. Its nature is not only festive, but it is also associated with entertainment, since it will be commonly advertised as an external communications event which the media are invited to attend.

On a more global level, management teams use this method for purposes of:

– As we have said, external communication: The game, turned into "event", allows the organization to display, when the reversal day takes place, the uniqueness and boldness of the reversal to its clients, partners, and to journalists.

– Internal communication: The principle of a game in which a boss carries out one of his employees' tasks allows him to discuss the

8 We notice that this management process seems to be mainly used in the entertainment industry [NEW 12] and in service businesses [SAV 13], two sectors that, as we have seen, made it their priority to develop principles of fun management.

democratization of hierarchies and closeness relationships, as well as to talk about cooperation, listening, and sharing. This is also the case for teams in tense situations (for example designers and sales and marking executives in an advertising agency, or field agents and administrators of a local authority), who are led by the job reversal to exchange views about everyone's experience with understanding.

– Learning: Learning is not related to what we have said about a job but, rather, it pertains to everyone's role and place in the process of providing a service. The goal consists of making taking over and business communications smoother processes by limiting conflicts. The objective is to allow, by literally impersonating someone else, to adopt another point of view supposedly conditioned by the kind of activity and specific constraints encountered.

– Motivational strategy: Recognizing the nature of work is one of the key concepts of reversal days. The aim is to strengthen the teams' motivation by bringing team members or line management closer or, as we will be told, to limit turnover.

2.2.5. *Business theater*

Imported from Quebec in the 1980s, business theater allows businesses to operate in several ways:

– Using a unique method, similar to coaching, that allows groups of staff members to take the floor by employing techniques used in drama.

– Staging sketches in order to animate large seminars or conventions. Issues that the company directors have chosen to consider are played. Speeches are interspersed with small scenes, in most cases in comedic tones, or an "actual" piece that helps employees think about life within the company and given topics like psychosocial risks, security, discrimination, burn out, etc. is staged. One way these sketches work is by urging, once they are over, the audience to react to what has been staged in order to encourage employees to engage in a debate.

– Using theater for "real-life scenarios" during training. The trainer is in this case joined by an actor who will perform with him certain scenes that introduce the topic and then some other scenes linked to training issues. Those who take part can often be encouraged to write their own sketch afterwards, under the supervision of the trainer and the actor, or to perform together with the actor, while the trainer takes note of the elements staged and encourages everyone involved to comment on the scene staged. They may also be encouraged to perform it again after making some changes. The situation can develop in two ways: either the trainees play the whole sketch with the actor or they are encouraged by the trainer to play a particular scene along the lines of: what do you think of this character's reaction? How do you think he should react? What should he say or do? They may also be encouraged to ask the actor to play certain scenes again in a specific way.

Business theater is supposed to accomplish three specific things: it achieves communication and organization goals, and it drives change [LES 13]. In all these cases, the sketches are written beforehand after views have been exchanged with the training sponsor and two to three people, always appointed by him – as far as we know. Within the company, the time allotted to gather information about these operations is very limited. Besides, more general material (about different topics considered, which are common in a business environment: discriminations, psychosocial risks, disabilities, etc.) and other operations that have already been conducted in other businesses are also used to structure training. Its content, which is standardized, will be "embellished" by anecdotes and features of characters inspired by the accounts of the individuals consulted. Thus, trainers can refer to scenarios and information about recurrent training issues which allow them to be responsive to demand and their business constraints. Issues are considered more in relation to their frequency (in terms of attitude and behaviors) than the specific features of their development and effects.

2.2.6. *Serious games*

> "A computer application that attempts to combine coherently serious aspects (Serious) in a neither exhaustive nor exclusive manner, with instruction, learning, communication or information, assorted with the playing aspect of video games (Game). This association aims to be different from mere entertainment" [ALV 07, p. 9].

The term "serious game" was introduced in its present-day meaning in 1999 by a consultant, Benjamin Sawyer, and a researcher, David Rejeski, who employ it to define “a serious use of video games”, “for more than entertainment purposes” [ALV 07]. SG are ranked first of all in relation to the three functions listed by Alvarez and Djaouti. Most of the time, these functions are combined to different degrees:

– “Spreading a message: SG aim to spread one or more messages, which can be of four different kinds: educational (like edugames), instructive (like newsgames), persuasive (such as advergames) and subjective (such as militant games and art games). The same game can combine different kinds of messages.

– Providing a form of entertainment: SG aim to enhance the player’s cognitive or physical abilities (for example exergames).

– Favoring data exchange: SG are designed to favor data exchange (like datagames) between players or between the game’s distributor and players” [ALV 12, p. 23].

We can see that the objectives and principles of SG occasionally match those of other training methods (business theater, role-playing, etc.) when they are part of an immersive project that assumes that being able to participate stimulates engagement, memory and learning. However, SG differ from other non-digital games in one aspect: the constraints imposed by game design on the range of behaviors of the players and those interacting with them.

Even if trainers, during a session, can play a guiding role and prevent players from taking a certain path, the discussion about the game, its rules and the precise expectations of players affect the development of the training process and reorient its content, which turns out to be in part repetitive.

Among the several ways SG can be used (education, defense, healthcare, advertising, etc.), we can also find SG designed for business training. The scope of these SG is extremely wide for those businesses who invest massively into them: sales, telemarketing, finance, security, teaching software designed for clients and internal use, etc. We will focus here on those used for management and lacking any marketing goals, even if the advertising and marketing aspect is evidently always relevant to the business plan, as is the case, for example, for a recruitment SG developed to improve the reputation of the company in the eyes of several applicants and in the media.

Edugames (used for instructive purposes) and exergames (involving training) are two of the categories that include SG designed for management. The significance of SG, apart from the fact that people can play them on their own – sometimes from home – and its immediate feedback element (evaluation elements), lies in its ability to produce performance statistical data, which other kinds of games struggle to achieve. If we wanted to present the main SG used in business management, we could mention:

– Recruitment SG: they are designed for different purposes and consist of identifying the most interesting profiles for the company as well as creating a "pool" of potential applicants. The latter can obtain a job, an internship, and sometimes remuneration, as was the case for the *Ace manager* SG used by BNP Paris and designed for finance students. The universes of a recruitment game are often immersive and have applicants face clients, managers, partners, etc., while also testing their knowledge. Reactions, behaviors in the virtual space, interactions, and game time all affect the assessment. Thus, the game conveys the values upheld by the company concretely and not in a didactic way through the choices made in terms of game design. It is also used to shape certain techniques within the company as well as to get players accustomed to the way the organization works or make them aware of its professions:

> "A curated game experience can test applicants while also telling them stories that represent the company. This allows us to assess applicants from a perspective not strictly focused on their professional skills, while also presenting the business to them from the inside and in a specific manner. Therefore, applicants can find out that they are getting the right idea about the company and assess if they see themselves working for it for several years".
>
> In this project, if game designers deal with the actual game (how does it work? And, especially, which of the interactions that have been proposed will not change the purpose of the game? How will the information conveyed by the game be structured? What will the game look like? etc.), they also collaborate with people employed in different trades who are in

> charge of the serious content. Thus, if the game proposed is a role game (applicants embody one of the employees of the company, which leads them to take decisions in scenarios that derive from actual experiences and are proposed by the several professionals working for the company), it incorporates certain situations, most of the time MCQ's (multiple choice questions), which are themselves determined by psychologists and evaluation experts.
>
> A game of this kind aims to make the application screening process easier for recruiters, who receive about a million of them per year. A total of 70,000 people out of the 100,000 expected played the first version of the L'Oréal game (2010–2011). It allows the recruitment of 185 individuals out of the 500 expected. These differences resulted from a game that was considered too long by the players and from very difficult evaluation questionnaires about professional topics. The second version of the game was played more (120,000 students in the year 2012–2013) for the same recruitment figure".
>
> *Excerpt from Interfaces numériques* – no. 3/2014.

– Cooperation SG: some SG are developed as part of a network or in multiplayer mode. Faced with an imaginary topic, i.e. solving a given problem in a virtual world, players are urged to cooperate and exchange views in order to tutor each other, for example. These games can include pretext bound-universe games, the presentation of regular tests, virtual universes where a collective operation, such as that of the project SecuReVi ("*Sécurité et réalité virtuelle*" or "Virtual security and reality") designed to simulate the coordinated and regulated action of firemen on a Seveso site [QUE 03], can be simulated. In this case, training consists of the sharing of everyone's skills and roles. "The social environment is structured and each member knows its roles and those of its partners. The interactions between the team members are also structured and arranged by means of a procedure known by all members. The generic organizational model is derived to formalize this concept of team. (…) On the other hand, the environment being dynamic, agents can sometimes need to adapt the scenario to the environment. Procedures then have a semantic representation so that agents can reason above. Procedures describe interactions between agents in an optimal case, and leave to the agent the responsibility to build implicit plans (not clarified in the procedure) considered natural within an applicative situation" [QUE 03, p. 170]. Another way these games work consists of creating teams of players and asking them to combine their skills in order to collectively solve a strategy issue for the organization.

– Change management SG: these range from games favoring a so-called learning "culture" (making co-workers open to innovation, improving their ability to adapt) to those that allow us to learn as well as test new methods without direct consequences for the organization, since it is "not for real", and thus to prepare for them, while also developing the expected skills. These games are based on a human (how will the player's interactions react to change?), technical, and financial dimension. "Errors" and "achievements" are assessed and players are regularly told about how the decisions they took have been evaluated.

These games are commonly based on several principles:

– Knowledge of the company environment and its present and future working principles.

– The development of skills through immersion in "realistic" situations. Players must make choices, take operational decisions, and take concrete action which will later inform progress reports.

– Communication between players and/or knowledge about what the higher and lower roles of the organization consist of compared to one's own.

– The integration of newcomers.

– Career management SG: Combining training SG with specific activities that contribute to advancing the career of a company's co-workers, GRH SG mainly deal with the employees' career and their assessment (year-end report, etc.). For example, the APEC reported that: "EDF employs a SG to prepare yearly performance reviews. This module aimed to train managers to carry out yearly performance reviews. Afterwards, it has developed a version of it designed for employees, as an exercise, so that they could train themselves for these reviews. (…) Renault Trucks, for example, does not use [it] for training purposes, employing it instead to assess training needs. Sellers, after taking a test, must use their avatar to cope with six different sales scenarios. Then, with the results in their hands, training managers advise a type of training" (Apec, 27/09/2012).

– Project management SG: They are based on the same principles of change management. These immersive and sometimes cooperative SG are centered on issues of global knowledge of the context and project, the diversity of skills, the assessment of the impacts of decisions, and human, technical, financial, and temporal aspects. This kind of game is commonly based on a classification of the problems most often encountered in project management. Several elements play a part in the game: team, line management, and steering committees.

Discrimination, psychosocial risks, crisis management, social and environmental responsibility all represent other management issues related to SG.

According to Courbet *et al.* [COU 13], SG facilitate learning in relation to three general features: the "reduction of the psychological stakes" linked to testing choices and behaviors (namely the trial-and-error methods offered by a SG); the gameful dimension and "game challenges" make learning less tedious; finally, the personalization process – thanks to roles and avatars, for example – implemented in SG programs tends to engage players and, through the personalized feedback provided by the game, to enhance "two types of meta-cognition: self-efficacy[9] and self-management skills[10]" [COU 13, p. 10].

Unlike role-playing, we can also consider that the fact that SG can be played through the mere man–machine relationship and/or the use of an avatar does not allow others to find out about one's possibly unsuccessful attempts. Testing, carried out in private and under the guise of an avatar (even if in role-playing the role played can have a similar function), may tend to make the game less challenging. Thus we assume that the stress levels associated with SGsituations are much lower than those observed when social games that require people in the flesh take place[11] Conversely, the aspect that Courbet *et al.* underline about a "role model" (a nice person who guides the adolescent or child in a game about learning good health practices) may be assumed to become more marked when games are physically played with other people. Trainers and players reputed to be

9 "Someone's belief in his or her ability to use the resources required to master certain situations" [PAT 05].

10 "Ability to govern in a satisfactory way one's own behaviors and their causes, so as to reach personal goals" [PAT 05].

11 Bertille Patin, while studying role-playing in training practices designed for adults, notices "the shame, the unease – or, on the contrary, the pleasure – felt at the end of certain more or less successful training sessions" [PAT 05, p. 163].

efficient become role models all the more often as this role is observed and agreed upon within a group during the game. What is unique about a group, in a game context, is its ability to cooperatively create representations and solutions by clarifying, showing, and breaking down the ongoing actions and decisions. In this respect, Bertille Patin points out that "since it favors development, a group makes learning easier" [PAT 05, p. 167].

2.2.7. *Traditional games (board games, cards, Kapla and Lego blocks, murder party, etc.) introduced in work contexts*

One of the games commonly proposed by advertising agencies to companies and designed for internal communication teambuilding is "murder parties". A form of role-playing that consists of carrying out an investigation, its board game counterpart is the famous Cluedo. Its main goal is entertainment and companies use it to allow their employees to get to know one another and also cooperate while playing together, without forgetting the organization component inherent to the game, which allows trainers, group leaders, and "game masters" to fulfill their role and adjust, control, and comment on everyone's actions during the game.

Murder parties may be occasionally designed for external communication purposes. This is the experience reported by Sébastien Célerin and Franck Plasse, "storytelling" consultants and authors of "*Gamification – methods, issues, and case studies of gameful communication*":

> "We have joined the Oise Departmental Archives to determine and develop a gamified way of discovering this place and the types of jobs that allow it to function during the very renowned (yearly) European Heritage Days. A place of this kind does not attract a crowd. Archives are generally regarded as boring and dusty places by a fairly substantial number of our fellow citizens, with the exception, perhaps, of genealogy enthusiasts. However, the Archives gather, look after, and restore a large number of documents, some of which deal with topics that captivate people of all ages, like the exhibits of a court case.
>
> The archivist perceived the potential appeal of a case whose documents were kept by his team. The history could have been drawn from the best detective stories. He also figured that the general public would find it interesting to discover this case

> through a game. I suggested he draw his inspiration from murder party methods, namely a meeting during which the guests play certain roles in an investigation, with the goal of piecing together imaginary facts that are supposed to have taken place before the game meeting started. In our case, it was a matter of asking the public to reconstruct the past investigation and his team to present the facts as they had been perceived by those involved at the time and, naturally, in relation to the official reports and archived evidence. Neither had to play a role like in drama or a murder party. On the other hand, they interacted in systematic ways (visitor-investigator and professional-helper) in a role-playing game on a well-known topic: discovering a crime scene and developing hypotheses on the chain of past events. In that case, there was no need to use any specific technology. What had to be conceived was a non-linear action, a space in which elements would be presented and where the professionals could shed light on them, the reason behind their presence there, and the way we managed to get them".
>
> *[excerpt drawn from Interfaces numériques, 3/3, 2014]*

We can notice again, as with reversal days and recruitment SG, how using a game can make a business more attractive, in particular from an outside perspective. Games promote the approaches of an organization since they allow us to bridge the aforementioned gap between the traditional opposition work/leisure time, and they update the "modern" qualities of the company while also showing an attitude that "does not take itself seriously". Here we find again the notion of "soft management" and its principles operating a "laicization" – understood as a "gamification" – of work.

Gameful teambuilding methods that employ Lego or Kapla blocks games follow the same logic. They may perturb some of the participants, who are hesitant about what they regard as a child-like practice. Their immediate goals are the knowledge among colleagues and cooperation, even if they obliquely aim at assessing how teams work: who has the right ideas, who slows down the team, who mediates in case of conflict, etc., are all questions asked by those who implement these games and the leaders and observers of their development. Once again, the role played by the collective "debriefing" of everyone's discussions and actions allows us to smoothly reintroduce the right practices in the shape of collective co-ordination.

The introduction of so-called "traditional" games into business contexts may also aim to vary the types of learning (this is also the case for business theater) and put forward new didactic methods. Using a transformed version of "Trivial Pursuit" for technical issues or testing the right/wrong sequences of actions on a board game follow this business logic involving learning and, above all, the creation of methods used to exclude participants with an attitude supposed to be potentially "passive".

These traditional games have explicit aims (even though they may be combined with indirect objectives like the evaluation of personal and team dynamics, etc.): finding a criminal, answering as many questions as possible, building the highest tower, etc. They also represent social games that encourage collective dynamics which, according to the kind of game, may involve competition and/or cooperation to reach these objectives. Players are either partners or rivals and are all involved in the evaluation of everyone in a way that is congruent to contemporary management practices (360° review, internal customer methods, etc.). The game allows us, at a given time, to reveal the actions of each player to another and encourages an individual to evaluate, reward, or punish him or her. Board or card games are also determined by rules, which are less flexible than those allowed when improvising in role-playing. These rules, which are explained at a given time, structure the development of the game. At the same time, the classic model from which these games are inspired (card or board games being commonly used in France) easily instructs the participants about progress issues, questions-and-answers, the higher bid principle, types of cards, etc. These games, finally, often include a random element (picking a card, rolling the dice) which makes the competition between players and teams less challenging, within the context of the games, according to a winner-loser principle ("There must be an element of randomness so that the losers can hide behind the chance factor" – says B, developer and publisher of business games). These games are generally compatible with the information required to determine actions in a game (sorting, response).

Consequently, these games can handle hybrid forms: competition and role, competition and chance, simulation and chance. What characterizes them is that they simultaneously represent a structure and an object (Lego blocks, board games, cards), they can be altered to suit a business content, and they can be organized in "matches".

2.3. On the field of games in business: simulation and role-playing games

As we have said, we have focused our research on simulation games and, in particular, on games involving role-playing used in management contexts. This choice includes a heterogeneous set of games and gaming contexts. Our analysis also incorporates real-life scenarios used in business training as well as role reversal games between colleagues and bosses or employees, business theater, and media processes where only a director decides to take his or her coworkers' job for a given period of time for work experience purposes.

What these games have in common is that they all belong to Caillois' category of "mimicry" and "make as if" or – and we will analyze the difference between these two aspects – "not for real". They all involve playing a role, whether someone else's (the other's role) or what might be one's own in a certain scenario (projection role). Finally, they all share the fact that they pertain to management: they may involve managers as well as have to do with management issues (evaluation, recruitment, teambuilding, etc.) that affect managers.

Due to the variety of simulation games observed, our corpus is not homogeneous. In terms of "living" material, it is made up of several aspects that can be observed directly: role-playing (consulting firms), business theater (a day at the hospital, another in a research center) and reversal days (advertising agencies, a hotel chain). This study is also based on a series of interviews with several individuals involved in these games: we have interviewed 16 business theater and role-playing professionals and 28 managers and senior executives that involve their teams in games for management purposes. These interviews cover different sectors, from the hotel industry, to business, the restaurant industry, administrations, etc.

As for indirect material, it is based on cinematic elements, like the rushes of documentaries on recruitment games and a teambuilding game (which amount to 16 h of recorded film) or "raw" films on two role-playing games in a consulting firm (10 h of film), images from a reversal day (1:15 h), as well as a documentary on a recruitment game[12] and excerpts from a

12 *La Gueule de l'emploi*, Didier Cross, 2011.

documentary on contemporary management practices[13]. We also include, as other cinematic elements, which this time are decidedly staged, the six episodes of the TV series "*Vis ma vie – mon patron à ma place*" ("Try my life – my takes my place") as well as six episodes of the more recent *Patron incognito* ("Undercover Boss"), a series in which a company director plays the role of his employees.

In light of the elements that make up the topic on which we carried out our research, games are certainly studied from a management perspective. From the creation of games in response to the requests of work organizations to the managers' jobs they involve, we analyze games as used in a management context. Our aim – this is one of the boundaries of our research – is not to question the effectiveness of games (there are several studies in business administration and learning sciences that deal with this topic) or the point of view of the employees involved in these games, which should also be analyzed in a different work. Across the themes tackled, the issue of our research pertains to the use of games as seen from a management perspective and the compatibility between the structure of these games (all simulation games) and, as it were, the game "work" expected by management. These expectations may be explicit (training, promoting interactions, coaching, analyzing methods, etc.), but they double up as the elements that games automatically involve, i.e. the perfomative dimension of games.

2.3.1. *The games analyzed*

2.3.1.1. *Reversal days*

Reversal days, the principles of which have already been described (section 2.2.4.), have been observed in two different contexts: a consulting firm (C) and a hotel chain (H). An exploratory study focused on 12 companies organizing reversal days, and included interviews with managers for two of them. A reversal day refers, as a term, to the principle of carnivalesque topsy-turviness, i.e. "the Feast of Fools": for a festive day only, a boss does the cleaning for company H while a developer becomes a CEO in firm C. Each employee – in the case of C – and a whole team of functional managers, in the case of H, are encouraged to swap places and roles with their colleagues. In a previous work [SAV 13], we showed the similarity between reversal days and carnival in relation to these

13 *La mise à mort du travail*, Jean-Robert Viallet, 2009.

apparently transgressive role-reversal games, in which directors sit in their employees' chairs. This correspondence is also based on the fact that this is an event that disrupts, if temporarily, the everyday reality of work. One of the functions of carnival is to denounce a routine. Hence, each participant does during a reversal day what he or she would not do on an everyday basis. His or her points of reference have shifted: a different place, job, way of organizing files, another work unit, and other tasks to carry out. A reversal day is also a playful interruption provoked by the role-playing involved in carnivalesque practices: in the two companies studied, a reversal day is a day in which workers swap clothes. Management at H wear cleaners or janitors' uniforms; C's CEO substitute for a day for once does not wear his piercing, shaves himself, and puts on a shirt; the saleswoman replacing the designer wears some sneakers and her son asks her if she is off work when she leaves in the morning.

Senior management find themselves carrying out the tasks of "low-level staffers", while those occupying the bottom positions of the hierarchy can, on a total reversal day, rise to top roles. However, as is the case for carnival, we can see nearly immediately that this reversal has nothing systematic about it. For example, men can dress up as women, as when one of C's employees goes to work wearing a skirt, high heels, and a low-cut dress to replace the receptionist. Staff also make men who have taken over women's roles dress up ("Oh aren't they pretty?" they say to the two managers who have taken over the accountants' jobs, "If I was a secretary, I would come to work in a skirt, but that won't cut it here" says L, the CEO substitute). Women do not dress up as men. In the company C, there are three kinds of dressing-up: involving gender, i.e. men dressing up as women, hierarchy, i.e. employees dressing up as bosses (wearing a shirt, nice shoes, close-shaven), or post, i.e. sales representatives dressing up as designers (or functional to operational in company H). The fact that women do not dress up as men, or bosses as employees, recalls the carnivalesque practice of hierarchical subversion: men are turned into women, a king is elected for a day among the people and his role will consist of an imitation of power. Who dominates becomes dominated and the dominated turned king irons a shirt exceptionally well for the occasion...We can ask ourselves some questions about what a comedic situation involving women dressed up as men would look like, or the mocking of employees, which is in all likeliness much less socially acceptable. A reversal day, unlike carnival, is not exempted from norms and decency. In firms C and H, those features that define someone's post, like a housekeeping uniform or the casual look of a designer, are still present in the dressing-up games[14]. In this respect, dressing-up is not exclusively related to a dimension of derision, but it highlights the function of carnival in relation to identity [PIE 88, PIE 05]. The specific attributes of typical jobs, while being overemphasized, are stressed as distinctive features on this day. Finally, in those games that involve power reversals, what is mocked is what it is possible to ridicule in virtue of the social privileges it enjoys: being a boss or a designer is socially well looked upon (moreover, the post of CEO and designer will turn out to be the most desirable when the process takes place).

Consequently, festive subversion follows authorized forms where management does not ape a low-staffer and a designer does not mock an accountant. What is despised about the job or status is not expressed or ridiculed. Women do not unsettle the established sexual hierarchy: they are the ones who are mocked.

However, this process differs from carnival (and gameful practices, according to some of the theorists we have come across) in one aspect: people do not necessarily take part of their own accord. This can be associated with the traditional notion of opposition between games and work, which Marie-Anne Dujarier underlines: "A game is an activity characterized by the fact that it is performed voluntarily for its immanent worth, i.e. the pleasure related to the elaboration, feeling, and production of meaning in a social environment" [DUJ 12, p. 92]. Reversal days, under the guise of games, actually involve a kind of work that includes game-related benefits linked to the dimension of "pleasure" and, consequently, associated with the engagement of the players. Both the hotel manager of H and the employees of C have obligations. Only the head office management of H are free – because they support the idea or they like the experience? To demonstrate to their colleagues and superiors that they are "committed" and open-minded? – to choose if they want to participate or not.

14 This explains how in firm H, bosses can dress up as employees because some uniforms are linked to certain functions. Nonetheless, we should point out that dressing-up takes place in the absence of those who wear them on a daily basis.

Another difference lies in the absence of consequences, at least in terms of material effects on reality (for example, the actual consequences of buying decisions when playing Monopoly, since gambling is no longer considered relevant to games in this respect) entailed by the game [DUJ 12], which is played "for nothing", unlike work. Even if it does not follow a logic based on short-term profitability (since there is no one replacing the players at work on that day), a reversal day involves management expectations of intercomprehension, but it also has consequences in terms of intervention on the work of others when it is management that "play" employees. The managers of C as well as H expect to learn something through the game, invalidating its definition as such, since most of the time games do not necessarily have to instruct as long as they have teaching potential.

Finally, carnival time does not include, in either of the two cases studied, anything ritual: it represents a unique event for H and it is unilaterally decided by C's CEO only when and how he deems it appropriate (even if that tends to happen on a yearly basis). If there is a festive element, it has an official nature and does not leave much room for the personal exchanges encouraged by any collective celebration [MON 01]: talking about topics unrelated to work, creating bonds according to one's affinity with others, choosing some coworkers to go drink something together, etc. In this respect, a reversal day is paradoxical: it employs the concepts of festivity and authentic relationships in a work environment allowed by the topsy-turviness that characterizes this specific day (being close, understanding the other as well as possible) and attempts to affect social bonds. However, it makes it look like a constraint: teambuilding for H and intercomprehension for C. In this management process, festivity is combined with constraint and business methods are at the center of the exchanges that are being encouraged. The personal aspects involved become institutional. Unlike other internal communication events, in this case the goal consists of consolidating work links, rather than sociability and links at work.

The organizers interviewed certainly describe this day in the best possible way. Some of the employees observed working for C also share their approval in different ways, from the most hesitant to the most vocal: "It's quite funny. At the end of the day it's quite funny. Yeah, it's funny" (L, the substitute CEO, does not sit still, gets up from his armchair, and seems quite hesitant and uncomfortable). "It's nice. It's a good initiative" says N, sales manager, "Reversal day is great (…) I find this idea amazing" says A, the project manager, "I find it great that I can find out more about other people's jobs, but at first I wasn't that much into it. Till this morning. I thought it was wacky" says S, designer intern replacing for a day a lawyer intern, "Last year I was a director. That allows us to experience posts higher up the ladder that we usually have no access to. There's other responsibilities (…) thus we can find out about constraints, the technical aspect, etc. (…) this allows us [in relation to the accountant job he has that day] to look into aspects we usually can't spend too much time on. And while we can't change their life, we can better ours, as a web studio. (…) This way we can see how that happens, to control what goes on" says J, director of C's Web branch).

Different kinds of assessments are made: agreement with the implicit norm that the CEO post is "nice", judging the event to be "nice" because it allows people to find out about other people's jobs or to gain insight into the work of other employees (for managers).

In firm C, as well as in H, different attitudes can be noticed during the event. What the managers at both H and C share is that they will actively, even intrusively, invest in the posts on that day. This will enable a hotel manager to become aware of the possible needs of his or her maids and do what he deems necessary to meet them: "Well, I made a comparison with the way I was employed (…) and I realized that day that we had to share a vacuum cleaner and that made me waste quite a bit of time, in the end it was a bit long (…). Since I experienced the activity for the day, when I got back home I asked other questions that maybe if that day hadn't taken place…"[15]. That will allow the managers of C to take direct action in relation to the accountants' overview reports and methods: "She's an accountant, we're operations manager so we can think about things concretely" (CEO). Posts can then be controlled and adjusted; we are far from the dimension of "unrule" that festivity is supposed to involve [DUV 73, p. 56]. As they are currently used, reversal days are still based on a hierarchical principle: in company C, only managers can carry out the employees' roles[16]; in firm H, managers are the only ones allowed to refuse to be replaced, they are not

15 She was helped in that respect by the authority of "truth" linked to the experience principle [JEA 08].

16 In an outstandingly devaluing relationship where only managers can do someone else's work but not vice versa [JEA 08].

briefed, they can lock their files and get involved in someone else's work. This attitude would be unacceptable for any other coworker: locking one's files would be seen as suspicious (while directors enjoy privacy rights), refusing to swap posts would be regarded as unwillingness to cooperate during a work experience, while getting involved in someone else's work would be perceived as a form of aggression. Finally, for the CEO of company C, this day represents an occasion to identify who plays along and who does not[17], and consequently to evaluate his coworkers.

If the low-staffers working for H who do not get involved in this activity are actually passive since they are left out, we can notice that the employees working for C are also passive in relation to their one-day job. No initiative is taken involving someone else's job. People are judged by the individual replaced according to the final result and, at the end of the day, everyone will admit how difficult they found the tasks of others.

Managers' interventions are legitimized[18], there is still scope for intervention, coworkers can still express themselves and even some symbolic places are maintained: during the reversal day observed, it is the "nuttiest of the company" (literal words) that will inherit the role of CEO. During the day, he will send an email about trading Whitmonday for a paid day off to the group of colleagues working for C. This email will cause a certain commotion within the organization and will make the CEO[19] laugh and who, at the end of the day, will all of a sudden turn the tables: "It took you some time to decide to stop brainstorming. But when it was a matter of getting a day off, you were quite quick. I hope you've learnt a lot".

We can assume that L, as the "nuttiest of the company", had to do something really dramatic, but that he represents no reason for the director of

17 "This allows us to see how people use their new function and also how they deal with certain kinds of management" (CEO).

18 24 Legitimacy is meant in its Bourdieusian sense: legitimacy as resulting from a social relationship between an authority that claims it while denying its structured nature and another that consents to assign it.

19 "I've just lost 20 000 Euros, in one go"
– With a smile to boot?
– Yes, I've been told it's about L's email
– How are you going to deal with this?
– I'll see. I'll see how credible L is as a director in the eyes of the people
– Is this the first time that something like this happens to you?
– Yes, it's the first time they've dared"

C to worry, taking into consideration past circumstances when "CEO" papers had been removed from the random draw to prevent unauthorized people from drawing them[20]. This situation is certainly a game of faces[21], namely a game based on expected courses of action, which are then supposed to be followed. C has its jester, a guarantee that transgression will be enacted… "not for real".

This subtle blend of features specific to carnivalesque games (reversal, simulation, laughter, transgression and grotesque through dressing-up, restored identities) and rationalization, as well as professional constraints (management expectations, preservation of hierarchies, for instance), makes a reversal day an ambivalent management practice where people pretend to play. What about the effects and limitations of this method in the companies considered?

2.3.1.2. *Role-reversals in reality TV "docudramas": "Vis ma vie – mon patron à ma place", "Patron incognito", and disorder reduction through games*

> "*Vis ma vie - mon patron à ma place*" ("Try my life – my boss taking my place") and "*Patron incognito*" ("Undercover boss") are two mainstream reality-TV programs based on a principle very similar to the one of reversal days, since it consists of power subversion: the CEO of a company takes over the role of his "low-staffers" for a given period of time. Some business reversal days, called "*Vis mon job*" ("Try my job") or "*Vis ma vie*" ("Try my life"), are actually named after the TV series "*Vis ma vie*". While not specifically a game, this still represents a form of role-playing: a company director takes over the role of one of his employees, i.e. someone in a dominant position swapping places with someone dominated (who will, in the case of "*Vis ma vie*", have the freedom to enjoy this). This role play also involves a job dimension: a bureaucratic director steps down to get experience in the field, i.e. manual and waiting jobs

20 We will learn during the day that the post of CEO may have been removed from the box when it was certain people's turn to draw lots: "I've removed my post when the person who had to draw lots was someone who's not really a team member when it comes to trust". Similarly, the managing director of the organization (the director of the web branch of the firm) will be allowed not to be replaced while he himself will replace someone else.

21 Namely, the "social positive value" that someone claims in an exchange [GOF 74].

> performed by professionals who are in most cases at the bottom of the social ladder. The two entertainment programs are based on dressing-up, the comedy resulting from seeing a director bossed around and, sometimes, the incompetence of those in power when they have to carry out "menial" tasks.

Vis ma vie "mon patron à ma place" is produced by Reservoir Prod. and broadcast on TF1[22] in 2006–2007 late at night. The episodes of this series were quite successful since some of them attracted up to more than 30% of the viewing public[23]. They would later be broadcast on NRJ12[24] in 2009. The principle of the general "Vis ma vie" series consists of having a celebrity interact with a worker and take over his or her job. Thus, comedians become intern bodyguards, while a reality-TV host takes part in the reality show by becoming an "urban climber", i.e. one of the workers who clean the windows of buildings.

The spin-off "Mon patron à ma place" is unique in that it shows the "actual" director of an "actual" company dealing with his employees' job. The media or cultural celebrity is replaced by the more anonymous, and yet more influential, figure of an economic actor.

The "Patron incognito" series, which is more recent, is produced by Endemol and it was broadcast primetime on M6[25] in 2012. It was quite successful and the first episode was viewed by 3.5 million people. This program is based on the fact that a boss will carry out different tasks in his business in disguise, under cover of filming the story of a long-term unemployed individual or, in more recent episodes, an intern. This version was adapted from the British series *Undercover Boss*, created in 2009, and still broadcasting new episodes on Channel 4 in 2012. Other versions were broadcast in the USA on CBS, as well as in Canada and Australia. The series was awarded an Emmy in the USA as best reality show in 2012 and 2013. It had been nominated every year since 2010.

22 A French public channel, with a huge television audience.

23 29 TF1 – The second episode of "*Mon patron à ma place*", which aired at 11:20 PM, attracted more than 1.68 million loyal viewers, i.e. 31.3% of the viewing public, which reaches 39.7% for females under 50.

24 A private channel (cable TV).

25 A French public channel.

As with business reversal days, corporate communication (internal as well as external, but also aimed at stockholders) is one of the targets of this initiative. The companies filmed advertise it on their communication media: O_2 posts an article about the program on their website homepage; UCAR highlights in its progress report the participation of its director to the program: "Our good business performances are the result of our dynamic franchisees, able to adapt to a difficult environment, as well as the growing power of the UCAR brand, especially after the positive consequences of the program "Patron incognito", where I had the honor to be the first guest in June 2012". As a BFM[26] journalist will comment, this is a "boss time" program.

From an internal perspective, these two programs are packed with references to the awareness of the unique nature of the other's job and, quite frequently, and the merit of those who perform it. "Getting field experience", as the company directors playing the game often say, is a way of bridging a gap inherent to the status and role of director: "It was a great opportunity to be able to spend time with the employees and factory workers, and to understand their life better" (said one of the bosses featured in "*Vis ma vie*"), "it is important to be in tune with what goes on in our business and to manage every post a little and see how things are going in the company" (ibid.), "Nowadays, people are closer in a business, which is an essential element for a service business like ours" (one of the directors of "*Patron incognito*" in BFM). The closeness issue is also an issue of reconciliation between management and salaried staff. Virginie Calmels, the president of Endemol France and deputy CEO of Endemol Monde, when interviewed by the magazine Stratégies during the Medef[27] summer academy, talked about the objectives of its production, namely *Undercover boss*, in these terms: "Everything that can improve the image of a director is positive, and this program contributes to it. People can realize that being a boss means first of all loving, managing, guiding, and making a business grow. Being close with employees is something moving: we can see the degree of empathy between directors and their employees".

The game involves closeness because it involves concrete interactions between field work and management, employees and directors. This kind of

26 A cable TV channel dedicated to (business) news.
27 The first French employer's federation.

closeness also relies on being in someone else's shoes and sharing a "common experience" that would ensue nearly automatically. The program offers an imaginary space, which is temporally defined and determined by an extremely inflexible plot structure (all episodes follow the same storyline rules), where directors are encouraged to play with their employees following preestablished dynamics which are known to all in the "*Vis ma vie*" series and still determined but unknown to the employees in the "*Patron incognito*" episodes. The closeness resulting from being able to take over someone else's role is presented as all the more spectular as the management and labor dimensions are introduced as extremely different, and even divided, worlds. These two programs present an aggrandized version of the directors from the very first minute. Their economic significance and even success is evaluated million after million. The two series show them while they pilot their own plane or sailing on their own boat, driving high-speed cars, or sitting in special offices, or leading a meeting. Amplification is the figure of rhetoric used: "Ucar comprises 260 branches all over France, its turnover is calculated at 70 million euros, its car fleet includes 12,000 cars, and it employs more than 500 people. The company is headed by Jean-Claude Puerto, 53 years old. He founded and has managed this low-cost car business for 10 years" (Ucar), "revenue of 78 million euros, 7,400 employees, personal assistance services, 140 branches, 27,000 people, business number one in France" (O_2), "Our great boss is the head of an empire: 22 casinos, more than 1,500 employees, 6 million visitors (...) Dynamic, a play-boy look, Benjamin Tranchant is at the helm of an actual empire" (Groupe Tranchant), "Well, Michel Morin, good evening. You're the president of an important chain of restaurants specialized in mussels. You're the head of 46 restaurants and more than a thousand employees with more than four million customers a year. What I say is, respect!" (Léon de Bruxelles).

This exaggeration, in contrast, brings about certain effects of dramatization linked to their change in status. The great boss "pretends" to step down from his seat at the top of the social ladder. Thus, the characters featuring on "*Patron incognito*" become long-time unemployed individuals for the first episodes, and then interns. Thus, Nicolas Riché, executive of Columbus Café, will be filmed while he puts down his Rolex and wears tennis shoes, a sweatshirt, and a long and dishevelled-looking wig. The TV series will make an exact copy of the scene in which the director casts off his status-quo clothes to wear those of an unemployed person, intern, or new recruit in the American version. In the episodes of "*Vis ma vie*", as well as in those of "*Patron incognito*", the jobs that directors will experience have nothing to do with team leader or middle management posts, but are most often at the bottom of the hierarchical structure of their business: waiters, cleaning ladies, car washers, mechanics, gardeners, etc. The first episodes of "*Patron incognito*" present people in precarious work situations: most of the employees filmed by the team have been working for less than a year, sometimes as little as a month, and are paid minimum wage when they manage to land a full-time job (for example, child carers working for O_2 who cannot be offered a 35 h contract).

The game mediates between two worlds that seem to differ in every possible way on a social and economic level. These universes do not seem to communicate, since it is necessary to "take over someone's role" in order to experience their reality, and because nearly all employees have never met their boss (which is actually a required condition in "*Patron incognito*" episodes so as not to reveal the "cover" (literal words) of the director). These programs and bosses hypothesize that this lack of communication results from the employees' secrecy and lies by omission, as they emphasize when they take over someone's role. This is what will enable the "*Patron incognito*" process, during which a boss in disguise will interact with some employees of his who have no idea that their boss is there, to take place. The CEO of Endemol will present the series in these terms: "At Endemol, we are also producing a series called '*Patron incognito*' in which a director is disguised so much that he is no longer recognizable, after which he moves within his company as a simple employee observing and discovering what other employees hide from him. Secrecy arouses first of all curiosity in viewers and then makes them excited as they are rapidly engaged by the story". The CEO of the production company will also say that "secrecy means power".

This incongruity between statuses and awareness of other people's identity will quite paradoxically be complemented by an equality principle made possible by the fact that a boss takes over the role of a paid trainee. In this case, the game is organized around the difference between the "for real" and the "not for real". Simulation and the "not for real" element belong to the fallen boss turned employee who experiences menial jobs in his company. However, lack of closeness, which will have to be rectified, ignorance, which has to be reduced[28], class gap, as well as the lies about the identity of the director that allow the monitoring process to take place without the employees' knowing (in the case of "*Patron incognito*") are all "for real" elements typical of programs that belong to reality TV.

The type of comments that directors make on their business and the gaps that need bridging within its organization shed light on how role-playing is used, i.e. as something with actual consequences on real life and with the ability to solve problems. In this case, the game represents a solution to a kind of disorder. By changing roles, which means – as these programs love saying all the time – "getting field experience", directors show their humbleness. They renounce their luxury and leave their comfortable environment to switch teams, literally, and experience someone else's post. The ambivalence of the game, which makes people interact with the opposite side, is in this case very significant in terms of social class and hierarchy. The opposition between the players is underlined by the constant distrustful speeches that directors make to their employees. Distrust has to do with information, which would actually be hindered by the social gap: "A CEO is someone with a very limited perspective on reality. Everyone wants to make your life better. You end up losing touch with the field. After a few years at the helm of a business, it is incredibly important to go find out the truth. You have to go look for it yourself, with your own eyes, and I guarantee you that the truth you're going to find out with your own eyes is not the same kind of truth you're told every day" (JC Puerto).

Taking over someone else's role entails entering a gameful and fictional dimension in which a boss becomes a mechanic. This game world will also provide information about a type of work in the actual world of which he is unaware. This is a point in common with the simulation principle, i.e. a space where the use of simulated experience leads to or trains for the real world. This other world, which is presented as tough, technical, or tiring in the episodes of "*Vis ma vie*", will require the acknowledgment of the professional qualities of the other, regardless of posts and functions performed. It will also be highlighted that a business is made of its

28 In the episodes of "*Patron incognito*", the UCAR CEO discovers during the shooting how one of his branches is doing. This is also what happens for the O_2 CEO. Lacking field experience is indirectly underlined by how they could have found this out without any game or disguise, either by visiting their branches or by encouraging feedback. Thus, it is middle management that is regarded as lacking since they obstruct this knowledge.

employees and not only of its directors: "I think that this experience will be positive for the employees, because it shows how nice their job is and how without them the business would not exist" (O_2 CEO in "*Patron incognito*").

However, it is mainly the game played for the most part by the director (he will be the one that dresses up and takes over someone else's role, even if some employees may play along and pretend to become their boss's director in the episodes of "Vis ma vie") who will receive all the praise.

The CEO is presented as a brave individual, ready to face the truth he is seeking. One of the directors filmed ("*Patron incognito*"), after seeing that one of his baristas is not friendly with all customers, will thus be allowed to say: "Today I've taken a punch, but I've taken it on the chin". The aforementioned truth, which is often mentioned in the two series, is not really delved into. There is a relationship of implicit equivalence between truth and experience or truth and "field", as if anything we are told can only be insufficient or false. The principle of immersion, in this case, seems to have scientific properties recognized by all: immersion leads to experience[29]. One of the paradoxes that characterize these methods is that in order to obtain this truth it is necessary to involve a fictional tool, i.e. games. The effects of time and media coverage are not analyzed. The field experience made by the director, which is very limited time-wise (he will spend a few hours on the job for each post presented) does not reproduce the elements specific to routine work, or jobs involving fatigue or changeability in relation to equipment, interactions, etc. Besides, filming (we could make the same remark about training courses where the employees' actions are displayed to the public by their peers) raises the problem of the media coverage of employees. As for the episodes of "*Patron incognito*", putting forward as a reason that the "disguise" of the director enables us to find out the "truth", could we think that employees will act spontaneously in front of the camera even if they are supposed to train an unemployed individual rather than their CEO? We notice during the program that criticisms are toned down and even withheld by employees, even when they are working in tough conditions (for example, washing a car outside in really cold weather): "Oh well, you see when it's cold this warms you up quite well, doesn't it? (…) physically, it's no walk in the park, we're not at the beach now, right?" (Hervé, rental car clerk).

29 This principle is often applied in sociological or ethnographic investigations, but it is regularly discussed on a methodological level…

Regardless of issues about truth, on the other hand, there is a lot of talk about inspection (in terms of respecting norms and procedures, equipment conformity, upkeep of premises, etc.) and checking. When making this experience, directors will rarely make comments on the consequences of their management policies (with the exception of the Hyundai CEO, who realizes that the rhythm expected can hardly be kept up). If it is acknowledged that certain jobs are difficult or physically demanding, it is in terms of recognizing the merit of those who have to perform them rather than reconsidering the work conditions. If we can identify certain adaptations to equipment (boots for car washers, training courses, raised beds for chambermaids to prevent backache), work rhythms and conditions are never analyzed from a critical perspective: it is always a matter of keeping them up rather than changing them. For the "*Patron incognito*" episodes considered, solutions very often have to do with training, relating the work to be done to the employees' skills.

Role-playing is combined with the comedic elements required by these programs, and musical accompaniment emphasizes funny moments. The "not for real" aspect related to roles is associated with a "for fun" factor. The audio cues used for comedic bits can be heard when the director finds out what he will have to face, i.e. cleaning toilets, carrying out bowel care on a septuagenarian or playing Obelix. The antagonism of the participants is also presented as "all in good fun" in the episodes of "*Vis ma vie*" and usually mixed with the narrator's comments (voice-over) – "…and his employees have decided not to forgive him anything", "In 48 hours our great director will live the life of his employees, who have decided to make him try everything. (…) No pity for the boss. And it's the right time for the employees to talk about things that anger the boss" – and the employees' remarks – "It's going to be quite pleasant to have my boss work for me", "I can't wait to see him suffer", "He's going to drown, call 911", "He doesn't really impress me, he's a man, just like me". Most of the time the atmosphere is "casual" and employees laugh with their boss about his role of managed manager.

However, the world of work represented here is shown to be quite tough. It is a world under pressure where professionals know stress, rhythms, inspections, dirt, precariousness, lack of solidarity. In the episodes of "*Patron incognito*", employees may be pitiless toward the (fake) unemployed individual that's being reintegrated into the world of work and plays a subaltern for a few hours. They internalize the management orders of rhythm and exclusion to reproduce them. Consequently, some evaluations are quite harsh: "sometimes he's a little clumsy and a bit of a numbskull", "he has to be more dynamic", "I don't think Ludo can be part of this company".

What the episodes of "Patron incognito" represent quite bluntly are those inspections made without the employees' knowing, as well as their trepidation and fear of their bosses when they are asked to have a talk in their office. Some are quite visibly stressed at the repeated dreaded prospect of being fired or punished. Besides, a substantial number of employees are noticeably uneasy when the boss reveals his actual identity and, consequently, the trick he has used. The spying principle of the series ("incognito", "a boss under cover") is met with the unease and silence of those who have been spied on.

The twofold structure that characterizes the game pits a reference framework, made of unequal relationships, against a fictional one in which a boss becomes, for a given period of time, a peer of the employees. In previous works [SAV 11], we have been able to associate the structure of the plot of "*Vis ma vie*" episodes with the plot of tales as studied by [PRO 70] and [GRE 80], without forgetting the structural analysis of myth put forward by [CLA 74]. In this respect, the episodes of "*Patron incognito*" are very similar to the "*Vis ma vie*" ones. According to Greimas, every myth or tale "shares a common trait [...]: the temporal dimension (...) is dichotomized into a before and an after "corresponding to a reversal of the situation" [GRE 81, p. 35]. The two series present a fixed storyline in which a daring boss sets out on a quest (for truth), is given powers to that end, provided by reality TV (becoming an employee, an intern, etc.), faces several challenges, and at the end reestablishes the initial order.

The fictional dimension of the game framework is consequently quite marked and clashes against the premise of experience as enabled by the patron's taking over the role of subaltern, as well as the categories in which producers class these programs. The aim of the two programs consists of presenting a "reality", since "*Patron incognito*" belongs to the genre of reality TV while "*Vis ma vie*" are "video documentaries" according to their producers.

As was the case for reversal days, we may be led to think that role reversal, just like in carnival, is well "outlined only in the most immediate and superficial symbols" [BER 94, p. 32]. As Michel Agier [AGI 00] suggests in his observations about carnivalesque practices, carnival involves distortion and detachment from reality rather than reversal or inversion. Thus, we could regard these gameful-carnivalesque forms as disconnecting institutional social relationships of domination. This adds to the detachment that results from replacing work relationships, converted into games, with festivity, even though "work is often presented as the exact opposite of games, which are the activities that most represent 'off-work time', leisure, and 'free time'. Besides, they are forbidden and even punishable during working hours" [DUJ 12, p. 90]. Everyday work and its inherent constraints, especially the hierarchical ones, are removed. This management initiative borrows from games their principle of "pretense" and the falseness on which they are based: "A player 'pretends' that objects and roles are true, while also being joyously relieved by the knowledge that it is not so" (Dujarier, 2012, p. 93). In this case, reality is completely inverted: we do "as if" we play, while we also do "as if" we work instead of someone else.

2.3.1.3. *Training involving games: business theater and "real-life scenarios"*

We will deal here with training days generally designed for groups of 6–12 people (15 at most) involving role-playing in two ways: business theater and "real-life scenario"[30] These types of training have been observed in different contexts: private businesses, administrations, healthcare. Direct and engaging observations have been made, but we have also had access to rushes and "raw" films of other types of training involving real-life scenarios. Finally, we have seen several business theater performances with promotional goals on different themes: burn out, resilience to change, harassment, management and teleworking, discrimination, etc. We have also been able to study, to underscore our point, a forum theater performance about France Télécom[31] and interview the two directors of the company Naje (*Nous n'abandonnerons jamais l'espoir*), a Theater of the Oppressed theater company, which produced this piece, about their professional practices.

30 Real-life scenarios are also used during training for business theater.

31 *Les impactés*, which can be accessed on Dailymotion: http://www.dailymotion.com/video/xn76edles-impactes-premiere-partie_fun.

Finally, we have studied two recruitment games involving role-playing (of salespeople and not managers), even though this data does not directly fit in with our topic.

Just like the arguments on which reversal days and reality TV programs were based, training involving role-playing relies on the soundness of the immersion principle. Playing the role of one of the actors present in one's work environment or "playing as if" are methods that refer to the fact of recreating the action and reaction conditions for one's professional practices. Here we find again Goffman's "technical reiterations", the aim of which is learning or testing through imitation or simulation. A technical iteration consists of a space designed for testing or training, according to Goffman, even though he highlights the distortions of the reference situation (the primary framework), which may be either simplified or made more complex in that context. In the types of learning studied, which make use of these techniques, a first observation has to do with the organization of the sketches staged or the real-life scenarios, which never last more than 10–15 min. This structure is justified by the reciprocal relationship between time allotted to simulation and time for commentary and discussion, which is called "debriefing". Thus, they consist of either "scenes" representing life on the job – "this incredibly trivial thing that reveals the whole system", in the words of an actor-trainer – or performances staged through a real-life scenario. These two methods are used to either show a "case" (a "delicate topic", as a trainer will tell us) that can be discussed afterwards (for example, a co-worker who does not want to accept certain changes that are being forced on him) and will represent the basis for further development (during business theater training sessions), or to challenge one's way of doing things and encourage people to try new things (business theater training and real-life scenarios). This creates a first difference between displaying, suggesting, and encouraging a debate (a scene of sexual discrimination or representing a patient and his or her family being mistreated in a hospital, etc.), and allowing people to test or demonstrate their professional practices or techniques (pulling a colleague back into line, convincing him to accept a project he does not want, etc.). In the report about the scenario, staged by the trainer actor, the goal consists of "writing something right" in order to "make people feel and think", while it becomes to "encourage discussion" during the debriefing phase. "The goal of the game is to recognize each other (...)

People say to themselves: someone is taking into account my reality". Trainers often refer to the cathartic factor inherent to theater because of its effects on memory and the desire to express oneself once the performance is over.

Real-life scenarios simultaneously involve the person engaged in the game and those who observe its development. For trainers, work is manyfold. For those who are made to perform, real-life scenarios "lead people to reflect, change, perceive with empathy which elements of a behavior will favor a constructive relationship or not" (trainer actor). The fact that the performer is faced by a trainer or comedian "prevents them from seeing themselves performing" since "they have to adapt to the person they are speaking to" (the trainer, in a way the game leader). The observers are encouraged to determine "what's wrong" and "what worked well" [GOF 74].

In itself, the situation played, when used to demonstrate a case or test something (we exclude from our analysis the practices of forum theater), determines and encapsulates the reference situations (the models). For example, people's precedents, which may affect their relationships and the nature of their interactions, or the systemic and cultural aspects of work organizations cannot be reproduced this way. Consequently, the main issue concerns problems related to the person rather than precedents, dynamics, or processes. The fictional element is related to the scenario, the interaction, or the scene played. There is no idle co-worker, played by the trainer or actor, even if his "scenario" is relevant for each consultancy team supervisor of the staff considered. The simulation is centered on a model – imaginary reference – which informs it. As François Daniellou says in relation to work simulations (even if the ones he considers are technological), every simulation involves three models: "a model of the human being, a model of the work system, and a model of the work that needs to be done" [DAN 07, p.78]. The scenes performed involve fictional situations which have, however, been noticed by some coworkers before training and some previous clients, who give shape to the trainer's experience and his array of "scenarios". These models, which trainers refer to, will either be, according to Daniellou, relative to "the normal situations considered by the creators" or they will "include an acknowledgement of the variability, the incidents, and the organization that will be implemented by the operators" [DAN 07, p. 78]. Real-life scenarios are structured by short written summaries that trainees will have to become acquainted with or be told about orally. The real world that serves as inspiration is always mentioned, even though modeling cannot ever enact it. In this respect, the issue does not consist of imitating the real world or the action that has to be performed, but in manipulating this "virtual reality or modeling" [SCH 99]. The unexpected linked to the occasionally radical improvisation of the actor or trainer, as well as the participant's creativity, is at the center of the discussions about improving these types of training: by staging real-life scenarios, they stimulate, improve responsiveness and the adaptability to contingency and what may happen.

We could say that these kinds of training are based on three principles: immersion, simulation, and model. Immersion, unlike the other two role-playing games previously studied, does not involve taking over someone's role or getting out into "the field" whatsoever. In this case, participants are encouraged to engage in a real-life scenario while also playing their own role. The principle of immersion can therefore be seen as an immersion in a "projected" specific work situation. As we can see, this involves a role since what is played is projective but also because what has to be played must follow specific rules (we will come back to the issue of "role" in the third part). A role is played in a "not for real" dimension and occasionally in a scenario of an actual situation. What the participant is asked is: "What would you do in a situation like this if you had this role in the organization?", given that, however, this situation is not real in itself since it represents a model, the collaborator played by the trainer/actor, or even a colleague, is not real[32]. and the problem we are encouraged to think about may not be faced. A projective dimension is inherent to simulation, which has been used and discussed a lot in ergonomics. A simulation game may be defined as a game that puts "a participant in a situation where (…), with the knowledge he has and according to the context he finds himself in, he must maximize his chances to win or minimize his probabilities of losing" [LEC 71, p. 261].

Participants play a "me" in front of their peers, representing a "me" which is not in an actual situation but is an ideal "professional-me" that must be efficient in the eyes of colleagues and provide the right answers. Real-life scenarios, hemmed in between two didactic training sessions, put forward most of the time a "model" of work, of the work system, and of the individual at work, if we want to use Daniellou's categories. These models

32 Consequently, it is someone we do not know and with whom we have not established any cooperation or share work experience.

will not be discussed since they are supposed to be (and are accepted as) representative of the real world or, to be more precise, credible[33]. In its most advanced versions, this "scenario" or model will give rise to what theater training professionals call "ready to play". During training sessions where participants are encouraged to "engage with a real-life scenario", only the role played by the participants is discussed, leaving out the modeling framework: relevance of the guidelines, behaviors, rules, and language represented, etc. In the cases studied, this role will be commented on according to management principles (right reactions / wrong reactions), since in all the cases observed, those participating in the training sessions have the same status and job. One of the trainers we interviewed told us that it is difficult and very uncommon (with the exception of teambuilding training) to organize training sessions that deal with the simulation of work and work relationships in a "real" team. Thus, training becomes aimed at managers or homogeneous groups. One of the blind spots of simulation involving role-playing games, unlike certain kinds of simulation carried out by ergonomists, is that it only very rarely combines those who participate in this activity and, in this respect, it does not reproduce the dynamics of work and cooperation or opposition at work by comparing the points of view specific to different statuses and workers. In this regard, there cannot be any "cross learning processes" leading to the "promotion of the application of learning systems" [BEG 04, p. 61] to a work situation reproduced in its entirety. The "model" issue becomes then more significant in that it does not allow the elements inherent to the status or job (in this case homogeneous) to be challenged, which would indeed happen if they had to be confroted with those typical of other workers. Consequently, what these methods call into question is only the individual actions and features specific to one job (expertise and social skills) instead of a simulated work activity. These types of training, consequently, focus on the job rules rather than the work situation, perceived as system and process. Another topic discussed

33 This is a difference from other types of role-playing games observed and used, for example, during recruitment sessions. During these sessions, groups of applicants for the post of salesperson or customer service representative (for the two sessions considered) are encouraged to play a role in situations that are not related to their future work ones (for example, they play a group of friends who are trying to agree on a holiday destination and each one of them has to justify the destination he or she has been assigned. The examiner's goal consists in observing everyone's reactions, their abilities to convince and negotiate, their relationship with the group, etc.

after the "real-life scenarios" have been enacted concerns the players' efficiency (their skills, how they deal with the unexpected, the relevance of their arguments, etc.). During this phase dedicated to commenting on roles in terms of respect of the rules and individual efficiency, everyone may be encouraged to suggest alternatives or talk about and describe his or her own situations, drawn from his or her real-life experience. In this case, participants have to ask themselves: "How do I play my situation?" (a trainer) and get ready to formalize their thoughts. In this respect, especially during the "debriefing" phase, participants are encouraged to "justify their chosen strategy", which underlines the pivotal element of "the relationship between the situation and the skills that have to be used" [PAS 06, MAY 06, VER 06]. Peers and the trainer will then approve (or not) these strategies by challenging the arguments used by the individual involved in the role-playing game. After playing, often before colleagues and trainers make comments on "the exercise", a participant/player explains what he or she has felt and thought while interacting with the trainer-actor. Players assess their own performance while also being led to defend their choices.

Occasionally, there are some initiatives involving a more complex relationship with the model, where trainees are encouraged to be part of the creation of this model. This is the case for certain methods used by business theater where the ultimate goal of training consists of having the participants "take charge" and create their own scenario themselves in order to stage it, sometimes in front of their superiors and, if not, for their colleagues. This also holds true for the forum theater practiced by the company Naje, which conceives (and stages) its performances by applying – making the most of all the time allotted to the process – a principle of involvement that concerns the development of storylines, the choice of the participants and the degree of reference to "the real world", i.e. what Daniellou calls the "type of simulation" [DAN 07]. This principle of limited use of preestablished "models" attempts to reproduce the fact that "everyone gets caught in a system that has to be described" (in the words of one of the co-founders of Naje). The goal consists then of "linking things together and seeing how they interact" (ibid.). Forum theater represents a process of collaboration on problems. This aspect is not present in the model on the practice of establishing certain "situations", or even breaking them into "scenarios" inherent to the simulation process even if, ultimately, it is a type of modeling that is presented during the performance. However, a performance will be presented as a fluid process that can be modified throughout since it allows certain scenes to be performed again according to the audience's wishes.

If these types of training are characterized by an analytical and reflective element (analyzing situations as well as objectifying one's own behaviors), using role-playing is supposed to provide a sensitive and emotional dimension to the problems tackled. As one of the trainers said, it is a matter of "speaking to the heart, making sense, and getting ready". We can find here elements of Jacob Moreno's psychodrama technique, which emphasizes the significance of physical and emotional aspects. A participant, while performing, literally becomes an actor and displays behaviors, attitudes, and emotions. Nonetheless, in the training sessions observed, which involve role-playing, we can see that emotions are "withheld", in the sense that those taking part in the "real-life scenarios" are not asked to reveal them. The emotion displayed may actually be stress when participants are encouraged to perform, but their interaction with the trainer is not introspective: it is a matter of showing what one would do in a specific situation. Consequently, it is the actor-trainer who introduces an emotional element: he may be led to play a crying character, or another who is aggressive or despondent, thus creating the conditions for the unsettling nature of his interaction. His work, involving the staging of sketches, the performances given during large seminars, or finally "real-life scenarios", consists of "finding a situation that will arouse an emotion and create things" (in the words of a trainer). Emotions involve the trainer of the "real-life scenarios" based on role-playing (many will insist on this aspect during our interviews), if not the participants. In this case, emotions are used for simulation purposes by having trainees face imaginary coworkers with a psychological life and who are quite unpredictable (unlike the more established relationships with "actual" coworkers). One of the participants will explain to us that it is important to assert the point of view of the opponent before the manager, who is trained: "We try to make different points of view tenable", for example by conjuring up the actual difficulties, which sometimes turn into a psychological burden, occasioned by a departmental reorganization. "We have to make up an ambivalent situation in which each party has his or her own point". This method engages the trainees' creativity in their attempts to answer and defend their position, or in the stances they assume after taking over their role. They are literally forced to improvise while faced with the unexpected, which leads them to employ "a lot of task-related skills" [PAS 06, MAY 06, VER 06].

2.4. ...Is it a game?

After our overview of game theories and our account of the methods observed in a work environment, we still have to find out if the topic our research focuses on, i.e. business role-playing, can be called a game or if it is only another symbol (among many) used to describe these practices. Clifford [GEE 12] underlined the significance in the contemporary social sciences of analogies with games (starting from the weight of Goffman's theory of theatrical performance and going back to the importance of the theories of Wittegenstein, Huizinga, as well as von Neumann and Morgestern, without forgetting Crozier and Friedberg). Geerz's critique focuses on the significance of these analogies, which are possibly invasive in light of their logical form and can encroach on other coexisting kinds of interpretation:

> "From Wittgenstein has come the notion of intentional action as "following a rule"; from Huizinga, of play as the paradigm form of collective life; from von Neumann and Morgenstern, of social behavior as a reciprocative maneuvering toward distributive payoffs. Taken together they conduce to a nervous and nervous-making style of interpretation in the social sciences that mixes a strong sense of the formal orderliness of things with an equally strong sense of the radical arbitrariness of that order: chessboard inevitability that could as well have been otherwise" [GEE 80].

Taking into account the relevance of this critique for the analysis of the social sphere, we could still argue that it is precisely this formalism specific to the game, theorized by researchers including those aforementioned, which is nowadays used to structure social interactions in learning or teambuilding processes in work organizations. If the scenes observed cannot obviously be regarded as "games" – in any case not systematically – from a *play* perspective (due to the marked variety and sometimes absence of play attitudes in the players), the issue certainly arises in relation to the *game* dimension, namely the structure of the game underlying the principle itself of gamification and its applications. A game – as "formal order" – is "automatically", as it were, a structuring principle (which constitutes the topic of the next section, part III, dedicated to the performativity of games). In relation to the cases studied, we will then claim that what we regard as a game is not the result of a formalization or figurative reading for analytical purposes. Besides, unlike an analogy that would limit the analysis, we consider these games useful for the explanation of social structures that games conveniently make inflexible: hierarchies, norms, strategic action and roles, just to name some of the most important ones.

Of the most significant characteristics of games is that they are shaped by rules, which can certainly be more or less strict in relation to whether we consider a cricket game or role-playing. In this respect, games represent a structuring principle of the interactions and meaning they have to be assigned: "The confused and tangled up laws of everyday life are replaced in this determined space and for this given period of time, by precise, arbitrary, and undisputable rules that must be accepted as such and ensure that the game takes place as it should" [CAI 67, p. 38]. Telling someone taking part in a role-playing game that he plays a manager who has to pull back into a line an experienced and valued colleague who, however, has a problematic relationship with another coworker, follows this logic. The situation, the protagonists, and the goal to reach are all determined, i.e. those rules that must be respected so that the game can take place. Caillois points that, in the case of simulation games (mimicry), if playing with dolls, for instance, does not necessitate formal rules, the principle of "doing as if" serves as one. We could add that the "doing as if" favors rules in the sense that, to simulate a given situation, we refer to the rules of the reference situation, which would be, for example, motherly behavior, caring, etc., for someone playing with dolls. As for the case of role-playing games that involve playing a manager, it will be the behaviors and norms linked to the status, to the post, to what we may be allowed to say or not in a company that make the "doing as if" acceptable. If the rules specific to the reference framework are violated or infringed, the "doing as if" breaks down and the principle itself of the simulation is invalidated. If I start playing an F1 pilot in a medieval role-playing game, it is certainly the game that collapses, since it becomes implausible (unless this becomes another game and the other players agree to it). The specific nature of the rules of simulation games is that they are borrowed from the reference situation, which shapes another feature of games mentioned by theoreticians: games create a separate space, both on a spatial and on a temporal level, which removes players from their everyday life. If this characteristic can be easily applied to a football game (except for when it is professional football, as Caillois reminds us), it must be adapted to simulation and role-playing games. First of all, playing a teacher and pupil game does not remove kids from their daily life strictly speaking and, second, the rules of the simulation game by definition prevent this game from being removed from other spaces or times of the players' life. Situations in simulation games are not isolated or contained in an "aside" space-time, even though they do take place in an actual space-time. Non-fictional simulation and role-playing games (such as playing the doctor, mom and dad, or a manager) are directly linked to other situations external to the game context. The delimiting feature of role-playing games can be seen, however, in the fact that there is a marked beginning or end and that a conference room in a company, for example, is used for simulation, namely a work scene that may as well stand for another, as we have seen with Bateson. The spatiotemporal dimension of games is a matter of convention linked to its application.

Besides the rules and the space-time specific to games, another features that theoreticians often mention is freedom (Huizinga and Caillois, for instance, refer to this notion). This characteristic, as we have seen, has been developed thanks to Gilles Brougère's research progressing toward what is called the "decision" criterion: "(…) games refer to a decision to play and getting involved in an activity with the possibility, at least virtual, of not engaging in it. (…) A game seems an activity that makes it possible to take a decision and, when the activity is mandatory and there is no way-out, it is possible to doubt whether this is actually a game, or at least one of the essential features of the game seems to disappear" [BRO 05, p. 51]. Besides, Roberte Hamayon [HAM 12] will point out the ethnocentric dimension of this concept of freedom by mentioning that in other cultures, for example among the Buryats, whom she has studied for a long time, games are compulsory even if, while they are being performed, we can find again the action and decision dimensions that characterize them: "Thus, if games are mandatory for the players (…), they always involve a margin of freedom in relation to their performance. Providing the possibility of manipulating, games also present themselves as an opportunity: to exert pressure for some, and to display their differences for others, overtly because of how obvious actions are" [HAM 12, p. 107].

The issue of the boundaries of the freedom to play is significant in relation to the context of "imposed" games, which is the one we are focusing on. In the games studied, the desire to play is not a necessary condition for their performance. However, during the game, players will participate to different degrees and, if we consider role-playing games, all players will be able to decide their game strategy, their actions, and what to say. Each player will choose certain aspects of the role he will take over, his ways of reacting to the others' role or, finally, his own choices in relation to the constraints resulting from the simulation: for example, how to be an "executive", which words to use in order to persuade or tone down a conflict, which actions to conceive to solve the problem he is faced with, etc. In the context of fictional games, these decisions will be characterized by another aspect that is often supposed (Freud, Winnicott, Henriot) to define a game: creativity. We may think that this issue, raised by psychoanalysts, ethologists, or educators, becomes problematic when it is associated with the requirement of "unproductiveness" put forward by Caillois and Huizinga. The latter focuses, for example, on the necessity of considering/locating games on a theoretical level "outside the sphere of necessity or material utility" [HUI 38, P. 187].

In terms of human and animal development, games are regarded as productive in that they produce links with others, as well as techniques concerning, for example, testing relationships, objects, or meaning. They are also thought to be linked to a purpose that Roberte Hamayon will call "productive" [HAM 12, p. 292], whether it consists of reaching a goal, winning the game in front of everyone or even, we could add, reaching the dizzy state desired as well as playing a good game of dice. An author like Delchambre [DEL 09] will call it "teleological" in that it "tends toward an end, which may be victory, profit, success, etc.". Consequently, the first feature of games, namely unproductiveness, is often challenged, since they are frequently regarded as a means of testing, which justifies in particular their use in a learning context. Delchambre will associate games with profit since they allow us "to learn about collective life", "to explore ways in which we can use and avail ourselves of objects or things", and "to claim for ourselves the symbolic resources (at once material and immaterial, linguistic and meta-linguistic) that allow us to develop an activity that we can call creative" [DEL 09].

In this respect, games will result in an evaluation, or at least in an assessment of the score, which could create a good game as well as a good player.

The decision criterion, as well as the creative scope of the game, is linked with the element of uncertainty inherent to any game. Uncertainty, in social games like a role-playing game or the simulation of a management job, will have to do with the action-related decisions and words spoken by the players. It will even more certainly concern the use of their creativity and skills when they are faced with preexisting regulations and norms which they can decide to play with or violate (as trainers regularly do in order to test the players they interact with). Finally, uncertainty may be mostly associated with the inventiveness of the players faced with a problem (like the trainer's violation of norms and regulations). As a consequence, the result of the game is "uncertain" in the sense that nothing is determined while the game takes place and when, during the debriefing phase, it is decided whether a player has been efficient or not, namely if he has been able to respect the rules of the game and the specific features of the profession or status considered (to remain a "good" manager), and if he has managed to react adequately to the mishaps that happened during the game.

Thus, we can claim that the creativity observed in these games is linked to the players' inventiveness in terms of combining the norms and rules inherent to the two contexts they are dealing with: the game context as well as the work one (training in a professional setting with their colleagues). They will therefore have to respond to the constraints and tests created by the trainer while also remaining "professional".

This twofold constraint – respecting the rules of the game and remaining professional – refers to Bateson's "double bind" [BAT 56] and will affect the criterion of frivolity that, according to Brougère, is inherent to games. In this situation of "double bind" – acting "not for real" (game) in a "for real" context (work) – it turns out that, in the words of Bateson, "No matter what a person does, he can't win" [BAT 56, p. 251]. What is done in most cases is then to choose to adjust and produce certain behaviors in the work context first, rather than the game one. A player refers then to the primary framework rather than the secondary one, using a defensive strategy against what may be regarded as a "falsification of signals" [BAT 56]. When games are used as evaluation tools, players can protect themselves by showing reserve. Observing game situations can show us that the work context and framework are never forgotten. Participants train, communicate, experiment, are coached…but do they play? This question reintroduces the dimension of *play* and playful attitude (Henriot) or *paidia* (Caillois), as opposed to *game* or *ludus* (Caillois). Although some of the participants enjoy themselves, make jokes, laugh about certain situations or surprises, and say they appreciate this type of training or event, this is not the case for everyone. Others merely play the role of participants and comply with the rules of the game proposed with seriousness: for example, playing a situation or writing a text. The person responsible for the game considered as *play* is, in a way, the trainer or consultant who, as we have seen, introduce unexpected situations, traps, challenges, or arbitrary tricks (for example, by regularly making jokes). However, we could say that the structure of the *game* works well: role and simulation rules, decisions, and uncertainty are all involved.

Another significant element that characterizes the structure of the *game*, i.e. its figurative dimension, is inherent to the role-playing and simulation games observed. With the exception of the reality TV games of "*Patron incognito*", all players are aware they are playing a game. They pretend as if it were real while knowing that it is not so. In the case of the "*Patron incognito*" episodes, only bosses, camera crews, and some accomplices, like the members of a player's board of directors, know that it is "not for real". Therefore, we can claim that in this case the game is lopsided due to the fact that the device is not known, which is a required condition for a game to actually be a game. With the exception of those situations in which the framework, which is presented as primary, is doctored (when the framework is "fabricated" and not "keyed", according to Goffman's terminology), the game context and its implicit element are known by all parties.

For the two games observed, the convergence between these two frameworks, the secondary game frame and the primary work frame – since games take place in a work context – makes interpreting this context ambiguous: is this a game or is this work? We could say that games, as productive, creative, built on uncertainty and based on rules, do not necessarily differ from work and that they share some of its features as an activity. Structurally, the games we have previously described simultaneously represent games and work processes: on oneself, on the situations, on the techniques that have to be used, etc. At the same time, we again find this discrepancy specific to games: what takes place in a game is not the same as what happens outside of this context. If only for the fact that, for example, the subaltern "pulled back into line" is not a subaltern but a colleague or trainer. On another level, what happens during a game does not correspond to what would exactly happen in the actual world. This holds true for the observations made during training sessions involving role-playing, the insubordination staged, or uncontrolled emotions. Once again, it should be pointed out that it is the actors (business theater) or trainers who in most cases keep the game within the boundaries of the professional framework. We can see once more the primacy of the professional framework over the game one. However, the latter manifests itself clearly through the ambiguousness of the game as a tool within the work organization and, especially, through the ambiguities associated with the fact of playing, working, simulating, being evaluated (by one's peers), training, displaying one's worth or weaknesses, etc. The "not really" aspect (biting, playing the doctor, pulling someone back into line) of the game, underlined by Bateson and Piette [BAT 77], works both ways: if work in the context of the *game* could not really be the kind of work it is supposed to simulate or represent, a game played in a work context could not really be the game it is supposed to be. Through ambiguity, in this case a game dimension similar to fiction is at work: "A player is not concerned with telling the truth. In his words, we can perceive the famous 'meaning jolt', the endless *Phantasieren*. What is most unsettling is that he may tell the truth, and we would never be completely sure of it" [HEN 89, p. 206].

Now that we are, all things considered, quite far from the notions of "rapture" and "enthusiasm" that Huizinga believed to be aroused by games and even from Csíkszentmihályi's[34] concept of "flow", we are actually fairly close to an idea of games as an arena [HUI 38] or competition games [CAI 67]. The "aspects" and adaptability of everyone are tested against the new rules introduced by the game through this figurative element inherent to the game between two frameworks (work and game). Testing can be of different kinds: "How should I behave with my (actual) boss who takes over my role for a few hours? How should I address him?", "How should I react in front of all my colleagues to this actor who is simulating a burnt-out coworker?", "What should I do today that, as a designer, I am supposed to deal with the sales department's customers?", "How to face the trainer's challenge about pulling back into line a defiant coworker?", etc.

These games involve adaptability and creative skills. The dimension created by the double bind aspect of games in work organizations (related to the issues, consequences, and goals of these games, but also the nature itself of the game as game) overtly requests the players to adapt to the double

34 "The satisfying, exhilarating feeling of creative accomplishment and heightened functioning" (quoted by [MCG 11, p. 35]).

framework they are faced with. In this respect, games can be considered to have a structural function: "The game produces events by means of a structure" [LÉV 62, p. 49]. Games work this way. What researchers and theoreticians have noticed about games is that they structure experience, which is what we still have to analyze.

3

Performativity of the Game: Games and the Structuring of Experience

3.1. From the reality of work to the fictionality of games

As we have seen, the "double frame" aspect that characterizes games can be expressed as:

– a reference frame that corresponds to the "actual situation";

– a fictional frame that corresponds to the "fictional situation".

It is the relationship between these two frames that constitutes games, which are formed, for example, by references to certain features, actions or behaviors; the game being simultaneously made up of elements "denoting" the characteristics, actions, or behaviors referred to.

Starting with Caillois, the importance of the "as if" element has been underlined. However, as Jean-Marie Schaeffer highlights, there are different kinds of "as if"[1] and the epistemic statuses of these "as if" must be regarded as distinct. In the context of the role-playing games observed in a professional environment, we will put forward three of them, which, according to us, are predominant:

– The "as if" of fiction understood and grasped by everyone. In this case, we are dealing with a common fictional structure, a "shared ludic pretense"

1 Jean-Marie Schaeffer also points out the distinction between modeling simulation and simulacrum: "based on different relationships between mimemes and imitated reality: a representation relationship in the former case, and a substitution relationship in the latter" [SCH 99, p. 94].

[SCH 99]. This type of pretense refers, for example, to the cops and robbers game: we pretend that you are a robber and I am a policeman, while we are both aware that neither of us is a policeman or a robber. We play the cops and robbers game. Along the same lines, we pretend that a CEO has become a janitor or carpenter working for one of his casinos and that the worker could, one day, be led to evaluate his boss's work. In this logic of pretense, the "truth" does not matter much: we can be a robber as well as a knight or a car washer while in real life we are the boss of an automotive group[2].

– The "as if" of simulation: We pretend that something is true, i.e. as if, for example, someone were a manager who had to evaluate a coworker or get him ready for a change. This is similar to the robbers and cops game in a way, since I can play a coworker while I am a manager. However, in this case, the reference to the "truth", meaning the (professional) world I am part of, is predominant. The "desire to transpose a reference situation" is pivotal [CAÏ 11, p. 86]. Simulation is modeling, according to Schaeffer's definition [SCH 99], in the sense that imitation is used to re-instantiate the professional behavior or situation imitated. The game explicitly aims to entertain or train and involves the manipulation of the scene thus re-instantiated. Simulation will rely on a clearly identified professional situation that will entail the creation of a role-playing game designed for training purposes.

– The "as if" of possibility: Brougère, drawing from Winnicott, refers to games as spaces of possibility, hence their usefulness in terms of learning. In its applications in companies, we can find an "as if" we had the right to test and make mistakes, but also as if everyone had the possibility of changing his or her place or the hierarchy itself (becoming an employee rather than a manager, working shoulder to shoulder with our boss, etc), as well as doing again what has already taken place. It is both a form of fiction (in the reality of work, unlike in a game dimension, it is not possible to redo or undo things constantly and spontaneously change one's status, while failing is frowned upon) and a mere common imaginary construction: we do not only play a game where a hotel manager becomes a chambermaid, we play the game of possibilities in the context itself of work and its organization: efficiency, hierarchy and strategies are all disrupted. Consequently, we do not play a game where we can become what we are not (which we know). We play a game where we can, in a game time and space designed to this end, do what

2 This is the case for the "*Vis ma vie*" (Try My Life) series of "*Mon patron à ma place*". In this paragraph, we used the CEO of the restaurant chain Tranchant and the company Hyundai as examples.

we are not allowed to do: stepping in and affecting norms, rules, and course of action. In this case, there is a higher degree of fictionalization, which does not involve the roles played or the scenes that have to be staged according to an explicit as well as unwritten agreement. Fictionalization, in this case, concerns the social organization of the professional world that fiction tests. This involves a game of possibilities similar to the transgression of the work order and its organization. In this respect, we reach a carnivalesque state of temporary disruption of the established order, which is an aspect we will come back to.

These three types of "as if" agree with Shaeffer's notion of "shared ludic pretense" [SCH 99]: the "as if" is shared by all the participants. They are informed by the presence of "conventional markers" [SCH 99] of ludic pretense: some aspects of speech, scenery or, if we want to consider another example, behavior reveal that we are in a fictional world and in a game dimension (this refers to Bateson's concept of "metacommunication"). As fiction, the games observed work in relation to three characteristics: "The idea that we are exempted from backing what we claim in fiction (…); the Batesonian notion of psychological frame; a definition that finally lets actors express themselves, both in their acts of fictional creation and in their disagreement about fictionality, its effects and uses" [CAÏ 11, p. 79].

This does not hold true for those situations that do not involve any sharing, as is the case for the "as if" of lying. In this case, we pretend that something is true even though we know it is not so. Here, the double frame will only exist for the liar, and not for the person being lied to. There are no convention markers of the game and pretense is no longer shared. It is in this respect that we discussed the game dimension of the "*Patron incognito*" series. If it represents a game for the boss, the TV crew, and the spectators, we may think that it is a form of deception for the employees, as long as the latter are not aware of the trickery. The fictional frame has a marginalizing nature and the scene reproduced is more similar to an opposition or split than a form of cooperation in the game (and consequently in the workplace). In Schaeffer's words: "[…] the situation of shared ludic pretense (…) requires mimetic techniques to stop being used in deeply manipulating relationships (preying or self-defense) and to be recycled by an intentional behavior that instead relies on a relationship based on transparency in terms of communication and trust between the person who produces the *mimemes* and the individual encouraged to go along with it" [SCH 99, p. 64].

3.1.1. *The ambiguity of games: from the "not for real" to the untruth*

Roberte Hamayon [HAM 12], as well as Elizabeth Belmas [BEL 06], points out how, in history, games as entertainment, or even passion, have been regarded as morally reprehensible in the Western Christian world. More specifically, if we consider those games that involve "roles", we could say that as soon as a representation is involved, there is suspicion of untruth or omission. The "as if" of the game and the representations it offers, for example in its theatrical forms, have consequently been long banished by religions. Historically, many of them criticized representations – that which represented something – by likening them to untruth. Roberte Hamayon writes about this topic: "Already blameworthy as imitation, representations are also criticized for the fact that they make the imaginary beings represented so realistic that they end up believing in their own reality. When human beings, through their physical behaviors, are thought to represent a being alive in another reality, rather than a fictional character, we no longer refer to representing or embodying and talk instead about being 'possessed'" [HAM 12, p. 66]. Hamayon, referring to theologians like Tertullian, mentions the connection between games and deception.

Representing literally means making something present. Thus, we discern the idea of substitution in a representation. An artifact ("artificial effect" from Latin factum, "effect", and ars, artis, "artificial") replaces something thought to be "natural" or "raw", regardless of the possibility of conceiving it as a social construct. An artificial effect substitutes what it represents, which is therefore absent and made to disappear, as it were, by its representation. This is the same logic that will justify the fact that a manager plays a certain role in an immersive context. Thus, these directors make it seem reasonable that, in order to find out in person about "reality" (a recurrent argument used for role reversal games), it is necessary to get out in the "field", which involves using role-playing games. It is commonly believed that these games do not involve the representation of this reality but, rather, that this reality is experienced because of roles. A director would omit, in this case, the part of his own representations used to create the roles linked to professions, statuses and types (being a worker, cleaning lady, etc.). However, if we follow this logic, a reality could not be experienced through its representation, starting with the experience as a director that those most concerned, i.e. those working "in the field", could make. It is shown right away that managers would experience a dulled version of this reality or would not experience it altogether if they did not directly embody the other's role.

In the different role-playing games observed in a professional context, replacing someone else by experiencing roles (rather than representations) is a multistage process:

– it may consist of an imitation of reality for simulation purposes (this is the case for games designed for professional coaching, or training focused on the actions that have to be performed in a work context or related to testing professional strategies or techniques, as we have observed in relation to training sessions for managers);

– it may also consist of a simulacrum, i.e. appearances that do not refer to any reality. In this case, we consider again as an example those entertainment programs like the "*Vis ma vie*" (Try My Life) or "*Patron incognito*" series, as well as reversal day events in companies during which a boss becomes, for a (brief) given period of time, a worker, a cleaning lady, etc., in front of the cameras in a program aimed at the general public or as an internal communications operation.

We may hypothesize that a simulacrum is more similar to lying in the sense that the manager's appearances seem genuine. We can mention two examples. In the media speeches and statements made during these events, the boss turned worker seems to be getting closer to workers while his experience only lasts a few hours before he returns to his office (which is also filmed in the final scene – reestablishing the initial hierarchy – of the reality TV programs analyzed). His initiative is called "brave" by some of his colleagues themselves. However, closeness is just for show, especially as we learn that several of his employees are meeting their boss for the first time in several years. Let us consider another example. In the aforementioned programs, as well as in internal communications operations based on this principle, managers are said to "experience" someone else's job. Is it actually possible to find out in a few hours what the job of a coworker consists of?[3] What do we know about the tiredness, weariness, numerous impositions and material or technical problems associated, for example, with the job of a cleaning lady if we only carry out her tasks for a few hours? However, managers will take steps to obtain this "knowledge" of

3 Moreover, how can we avoid belittling it, if we implicitly claim that we can find out about it in just a few hours?

someone else's job at the risk of using approaches that are not suitable for a situation that is literally distorted.

This type of lying, which is performed and staged, pertains to discourse and, in terms of business management or internal communications, we can regard this kind of game as a speech made by management: about closeness, sensitivity to the work conditions, unequal statuses, etc. It is this speech taking the form of an act that may or may not involve a "true-meaning-to-say" (Derrida), and it will establish the difference between an untruthful game and a truthful one. Thus, we can assume a game designed for learning purposes to be based on a "true-meaning-to-say," which is the desire to train (even though saying that this game is pleasant to all of the employees or, if we want to use Brougère's term again, "frivolous" and without consequences for the participants may be a lie).

What happens to a game in which a boss becomes a subaltern? Can a game where a boss plays a worker be seen as a lie? Not so much because it is suggested that the boss is actually a worker (with the exception of the "*Patron incognito*" episodes), but in relation to the assumptions about the experience made and the closeness allowed by the game? If we consider, in Derrida's terms, that "the contrary of the lie is neither truth nor reality but veracity or veridicity, truth saying, the true-meaning-to-say (…)" [DER 05, p. 46], how should the situation be defined? In a professional context, what corresponds to the "as if" of the game: a shared fiction or a lie, and even a farce? When, in order to "get field experience", as managers who play put it, they carry out the tasks of a worker, they present themselves as pseudo-workers. The term itself "pseudo" refers to two categories of things: "In Greek, *pseudos* can mean lie as well as falsehood, cunning, or mistake, and deception or fraud, as well as poetic invention (…)" [DER 05, p. 11].

Here, games seem to fit, on the one hand, into the category of falsehood (in the literal sense of the "not for real" aspect of the game) and, on the other hand, into the category of creation (which, as we have seen, characterizes games, according to Winnicott, for instance). The ambivalence inherent to games, which we have come across when analyzing its related theories, is now resurfacing.

3.1.2. *From double assertion to mediation*

When we consider games, especially those involving simulation or role-playing, we can see that they interact with the real world because of their double frame: a bite simulated by a game refers to a model of a bite but it does not constitute a bite (otherwise this would not be a game). Bateson will call this bite "fictional". Albert Piette underlines the importance of denial in the game frame considered: "What matters is actually the negative aspect of a game. In the act of pretending to argue, which means "not really" arguing, what matters more than the content itself is the impact of denial" [PIE 97, p. 40]. If, in a game, a manager pulls a coworker back into line, this does not actually represent the act of bringing him back into line (just as this act does not really involve his coworker). However, if the game turns out to consist of pulling someone back into line "not for real", it still involves "for real" the issue of bringing someone into line: "[...] there is this peculiarity about play, that with the word 'not' also goes the word 'really', apparently used with the same sort of ambiguity as the word 'not'" [BAT 71, p. 264].

A boss playing a worker is not really a worker, but, following this logic, he is not a non-worker either. Bateson [BAT 71] underlines how, therefore, we can say two things that are simultaneously true: "he is not a worker" and "he is a worker", while the frame used for the interpretation (actual/game) can be chosen by the observer. This represents a space between, intended as mediation. In fact, the boss playing the worker or a manager embodying a subaltern display (and put into words while the game is taking place) what they think this position and post consist of, which they express in terms of courage, tiredness, difficulty, expertise, etc. "Knowing how to work on the assembly-line is very tough", "1,500 packets have to be delivered every day, this is already an incredible amount of work while sometimes I might think they're not working hard enough" (excerpts from comments made at the end of episodes of "*Vie ma vie*" (Try My Life)). The worker played by the boss in this game makes a speech from a management perspective concerning the worker (the "real" one) and the relationship between boss and worker.

Thus, trainers refer to the mediation allowed by games and to catharsis: playing work relationships, and the tension involved in these relationships, allows us simultaneously to maintain perspective and to portray delicate issues. "Certain things will be said fairly bluntly while someone's actually laughing", "They are no longer in charge, as they usually are, of what goes on in the office of a company", "The more fictional the situation, the better we can delve into delicate issues and find out the truth", "The more emphasized the fiction, the more people can relax and accept to become involved" (trainers). It is following this logic that Nathalie Zaccaï Reyners starts her article "Jouer pour penser" ("Playing in order to think") with Oscar Wilde's aphorism: "Man is least himself when he talks in his own person. Give him a mask, and he will tell you the truth". For trainers, games allow us to think about the real world and create the right conditions to make this happen. The "mirror effect" of theater is often mentioned: "an actor's ability to magnify certain issues that may relate to us", "there's no game that comes out of the blue, there is always a quest for truth behind" (trainers). One of the tricks explained to us by trainers who make use of business theater consists mainly of starting by saying that the scenes staged are over the top and fictional. If they are presented as true or plausible, they will be criticized as being caricatures. If they are introduced as we explained, the participants will realize how similar they are to the real world.[4] It is a matter of likeness and echoes, and the debriefing about the scenes will be organized, according to one of the trainers, in two phases: "(1) What did they see? They talk about the "story" that has been staged for them. (2) Which aspects will remind them of something? Does that speak to them? Have they experienced anything similar?".

Likeness and echoes, however, are put into perspective all the time, which is allowed by the decontextualization of the game. One of the trainers, while people are talking and making comments on a scene that has been staged and their professional experiences, will regularly encourage them to speak by saying that "this isn't about us" (the workplace). In the game practices employed in the workplace, appearances are regarded as a means of telling the truth; the "not for real" serves "truth".

Sometimes, the relation between game and reality can be reversed, and some trainers emphasize the games of appearances and the "falsehood" inherent to relationships in the workplace: "They'll actually say 'yeah, yeah, fine' in the corridors. That's all fake!" (a trainer).

4 This is what trainers are constantly striving for in order to orient their training: "Understanding a need means being sure that we will get our message across to them, give them something, and be useful to them. We have to get what they need" (a trainer).

3.1.3. *The liminality of the game*

However, in the workplace, we may be led to think that people, because of the primary frame (the professional frame) and what it entails, to a certain extent pretend to play[5] and the game itself becomes pretense. A sentence along the lines of "this is a game, therefore this is not 'for real'" can be challenged in light of the power and subordination relationships in a company. Are we not being evaluated while we play? In this respect, Piette underlines how: "We face a double-reference-frame situation with double condradictory instructions (...) Globally, it is a type of communication in which something is asserted and denied at the same time" [PIE 97p. 42]. Thus, a work situation can be presented to us as simulatenously true and false:

– it is a work situation (you have to play the game and pretend);

– it is not a work situation (you can go ahead and see it is not for real).

There is "liminality" in what happens during the game. A game bite is still a bite. If we follow this logic, what can colleagues settle during their everyday interactions in the workplace by making use of the ambiguity of the game? Some debates we observed during the debriefing sessions for role-playing games temper the mere "not for real" aspect and assumed friendliness of the game. The managers' reflexes are not only tested during the role-playing game, they are constantly put to the test during the discussions that executives have about what they have played and observed. We abandon the game frame to go back to the professional one. "A game always has to get us further. I do not play the game if afterward I cannot analyze the behaviors adopted during the game" (a trainer). Even if a certain tactfulness is common, several participants will be challenged and will have to either prove their point (or game) or accept the criticism and recognize that their point is invalid. We will consider now, as an example, a part of the training that advisors undertake where they are encouraged to drive one of their colleagues to take action and become proactive in a difficult context (reduction in funding). The trainer playing the coworker is opposed to this.

5 For those who accept to play. In some cases, we have been able to note either blatant behaviors involving detachment from the game (which, however, did not prevent people from listening), such as using a tablet, writing e-mails, and reading the newspaper, or distant behaviors, like silence.

The advisor being trained ends up proposing some possible solutions. This is part of the debriefing:

– (one of the participants, who is a colleague of the advisor doing the role-playing): he got into the role of the collaborator, he has fallen for it. His role consisted of being of help, but only after someone else has asked for it;

– (another participant, laughing): you are going to find it tough, you are too kind;

– (the trainer): yes, the idea is that even when it comes to a detail or a trivial idea, we say "what's your solution?". We have to be able to involve the other. If not, this is called the "bin manager": a "bin manager" is someone who acts, takes decisions, and comes up with ideas instead of a coworker;

– (the participant, just after finishing the "debriefed" role-playing game): well, of course, you have to have the time;

– (the trainer): it is exactly this, what depresses the advisor is that he has to find a way of managing without having the time to manage".

The "actual" work conditions are used as a justification (lack of time) and the participant refers to the reference frame rather than the "not for real" element of the game. However, criticism is made within the game context, which creates the conditions for its expression.

Thus, through games, we can play a game of limits: "We are in a limbo (…) where we can reach the very boundaries of the game behavior in question, without actually entering the "class" of aggressive behaviors and risking facing the consequences related to it" [PIE 97, p. 142].

Therefore, this raises the issue of games as a vehicle for contradictory discourses: what does a game allow us to say in a professional frame, which could not be tolerated outside the game context? "Soft" evaluation, "reframing", and "requiring the 'right' ways of working" all have to do with the functions of games, regarded as effective tools.

3.1.4. *"Belief forged within immersion"*

It is important to set "fiction markers" between game and professional frames – a requirement pointed out by Piette – so that a game can actually be a game. Which are these markers: roles? Relationships? Degree of plausibility? Context? Action time? In the games used in professional contexts, the "limbo" distorts the message and blurs the markers. Acting, which feeds on plausibility, also contributes to that: "I gave in to what David (a participant) proposed. I followed my sensibility as an actor. If people get up my nose, I dig my heels in. I bet that if this is what that does to me, maybe it does that to other people as well since I'm a human being" (an actor)

Piette [PIE 97] identifies five of these markers, which, according to him, make "nonsense seep into the most serious matters". These markers are defined by a common trait, namely an excess of signifiers, and consist of: enumeration (Piette considers the example of a procession, but we could also imagine the listing of the different "types" of personalities in business training and the sequential nature of games in role-playing); the repetititon "of a gesture or a set of gestures" (playing a scene again, doing an exercise a second time); the amplification "of a gesture, setting, speech" (the over-the-top nature of the game played by certain actors who abandon themselves to emotions or to an antagonism which is often downplayed); oxyomoronic contradiction (the boss-worker or cleaning lady is a good example); and asyndeton, i.e. the omission of linking elements or, in other words, consecutive units of meaning that are unrelated or lack the coherence that they are supposed to have (for example causal effects). Roberte Hamayon explains that when the Buryats imitate animals, "their imitation is, as a matter of principle, selective and partial; in particular, it exempts the players from the consequences that the movements started would have for the animals' [HAM 12, p. 123]. During games, something may be started but not pursued any further, then be taken up again, maybe sequenced, and belongs, all things considered, to an "aside" space-time that separates it from "real life". This split often takes place in the games observed: the time allotted for the scene and real-life scenario and the time dedicated to didactic learning and debriefing are clearly distinct. During reversal days, the time allotted to the meeting at the end of the day marks the end of the role reversal, while the episodes of "*Mon patron à ma place*" constantly end in the office of the director after he has taken off his mask.

In general terms, it is order that seems to be undermined in a game. Meaning is disrupted by processes such as splitting, reiteration, excess, reversal, etc. Another marker specific to the games observed consists of the fact that, while in "the actual world" meaning is scattered around, in this case everything can be understood in terms of another relationship with excess. This may take different shapes, such as the explanation of everyone's role in the role-playing game: "Well, now we're going to play a role-playing game where a colleague is already facing some challenges since he's been brought back into line because he's always late", etc.

However, this fictional dimension, which characterizes games in particular, turns them into the ambiguous space we have talked about: while fictional, games are not imaginary. They require us to "play the game". Jean-Marie Schaeffer [SCH 99] quite rightly questions the effects of this fiction on reality: which aspects of behaviors, relationships, and representations does it alter? To this end, Schaeffer encourages us to call into question the different "immersion stances": from interpreting a form of fiction to people enacting it in a role-playing game through theater or role-playing games, for instance. What is summoned during the role-playing game will be different: when I am reading, I refer to my representations – when I play with someone else, I am forced to do what this other person introduces in the game. This compels me to adapt to be inventive and responsive, which represents a sales pitch for this kind of performance in a business context.

What is summoned in fiction may be very different from our perceptions, but, in a game that simulates work situations, this difference raises some problems, in particular when someone has to play with other colleagues. In this case, what is at stake is something else: this is not about the aesthetics of fiction (as Schaeffer regard it in terms of its autotelic importance: being a good or bad form of fiction) to grab people's attention; it is about the referential plausibility of what is simulated so as to train, inform or raise awareness (these terms are all used by trainers or advisers of business games) in companies.

If we use again Nathalie Zaccaï-Reneyer's phrase [ZAC 05], what is called into question is the "status of belief forged within immersion": "Thus, it is primarily the status of belief forged within immersion that will change according to whether the practical framework is serious (for example during learning) or gameful (for example when accepting forms of fiction" [ZAC 05]. In our case, where we consider combined frames (the professional and the game frame), there will be several "game contracts", as it were (just as we have a "writer– reader contract" for novels), established between a player and the game frame. The range of its varieties will correspond to a spectrum that goes from a form of fiction that does not aim to be plausible (interactions and reactions that seem unlikely in the "actual" work dimension, the comedic goal of some of the situations staged, etc.) to simulation (re-instantiation of the actual world) and, finally, to untruth (the untrue-meaning-to-say) that divides the game into players and those who are played.

3.2. A role to play

In a game, one of the main means of immersion is represented by players themselves. There is no game that does not involve a player, who is at the same time the subject and object of the game, which is all the truer for a social game where the other players are as much part of the game as they play it.

In games used in work situations, we can see a doubling up of the "roles":

– a player plays a first role that corresponds to Goffman's notion of "face". The participant knows he is playing a game in a professional context with a defined goal (relaxation, teambuilding, training, recruitment, etc.) and he adapts his "face" to the game. Similar or comparable games would not be played in the same way among friends;

– consequently, two roles are involved: the (socially expected) role of the professional and the (socially expected) role of the player;

– as well as the role specific to the kind of game proposed and the order to play it, according to an initial script, so as to comply with the rules of the game.

When we consider role-playing games, we can find the essence of games, the "as if" or the "not for real" of the social games played by animals or kids on which mainly ethologists and researchers in learning sciences have focused on. In another branch of the social sciences, Goffman's theory about the contexts of experience matches his previous analysis of the notion of role. According to Goffman, role is what we play in the different social spaces of our existence following a logic based on our repertoire of available roles. We can say that the other's role is part of the frame: it is one of the elements that will define the situation (Goffman's primary frame). Role, as a predictable construct, pertains to what is said about the situation (metacommunication) and allows us to get our bearings when we are faced with it. Conversely, the frame will affect the roles. It is the frame that will define the "courses of action" required in the situation encountered. In the game frame, I can, for example, react in certain ways which would not correspond to my behavior in the simulated situation, namely the reference situation. I can also emphasize certain emotions that I would normally glide over, I can pretend and set a trap, "not for real", to those who are interacting with me, which I would not do in other situations outside the game context to prevent them from losing face. However, the inclination to play or lack thereof in those who are interacting with me will affect my abilities and desires to pretend.

However, all things considered, I play in the sense that I have the possibility to explore and test certain margins of action allowed by the game, which in this case takes on its meaning of space in a mechanism – just like a game between two parts of a mechanism. These two parts, as we have seen, are represented in part by the reference to the actual world that I simulate and the fiction I enact while also knowing, following the example of Henriot and Brougère, that if the game is fictional, it is not imaginary. A business director who plays the role of an accountant during a reversal day, or a manager playing the role of a coworker that must be convinced to accept a change, is not imaginary, even if what is being played is a fiction.

First of all, I intend to analyze the issue of roles and afterward the relationships between role and fiction in the context of role-playing games. What does it mean to play a role-playing game in a social space like a company?

3.2.1. *Role understood as status*

Belgian sociologist Jacques Coenen-Huther, in a paper dedicated to the "steps and stops of social roles" (dating back to 2005), returns to two "historical" meanings of the concept of role: the first is inherited from Ralph Linton and relates to the aspect of status [LIN 36]. The other one appeared at around the same time (1936) in the work of sociologist and psychologist George Herbert Mead, one of the fathers of symbolic interactionism, who regards role as an interface, as it were, between an individual and his social environment.

If we look into the definition of role in its relation to status, the former, according to Linton, consists of the "implementation of the rights and the duties that are the constituent part of the status" [LIN 36, LIN 68, p. 139]. In the domain we are dealing with, and in relation to the games observed for this study, one of the business theater companies studied is called "A role to play". In particular, it offers training sessions to managers. One of the services it offers is named "Managing, a role to repeat". Often, during the interviews carried out at this company, the same notion of managerial behavior often recurs, and the Web sites of these companies display such slogans as "Our goal is to be at their side so that they can overcome their obstacles, offering them to step back and see the bigger picture, as well as to work concretely on their stance, behavior and communications skills". We may find in this objective the notion of the role that must be played, in a somewhat strict sense of the concept, as someone embodying a status. This approach to the notion of role is somehow related to the demand to be a manager, which is equivalent to the demand to fulfill one's role as such. As for role-playing games, this meaning or representation of role leads to training sessions where the propositions of a manager in a specific situation are successively tested, challenged, accepted or refused. For example, what a business theater enacts, through the performance of an actor-trainer and an actor, is hypotheses about people's behavior in the context of the first developments of a situation (a conflict, a negotiation, a recruitment process, etc.). The participants in the training session have to debate this performance and, in some cases, make their own proposition by suggesting new behaviors, new modes of interaction or verbal formulations in the scene that has been staged. What is played again from the initial staging may be done by individuals, and it will then be discussed in the group. It can also be reworked collectively by the group, who gives their instructions to the actor for playing the scene again. "The group ends up saying: you cannot do it like that" (a trainer). If they are invited to play a situation again, the participants are sometimes encouraged to keep elements previously defined as "good" and that have been listed on the flip-board by the trainer during the debriefings.

In those two modalities, there is a collective work of discussion, then the establishment of a consensus about what could be "a proper way of doing it" by people who are sharing the same status. This status can be linked to "profession" (such as medical staff in a hospital) or hierarchical functions (as in the case of management). In this case, we are close to the idea of "profession" as a (historical) form of social organization, put forward by Dubar and Tripier [DUB 98], who refer to "identity instructions". In the practices we are studying, we may also find the initial sense of profession, underlined by these same two authors, i.e. "the action to profess aloud one's own opinions or beliefs" [DUB 98, p. 4], like in a profession of faith. The work on roles performed in business theater companies allows us to put into words a professional practice, which, in relation to this type of learning, is verbalized and acted out (it is played). Here, the action is declarative and meant to be displayed – it gives rise to a performance – which backs up Goffman's theatrical metaphor about the repertoire of roles that we are led to control so as to be able to perform them in our social life. The roles embodied by the participants during the training sessions are performed in order to be discussed and reworked, or even corrected. Trainers say that "It allows people to reframe, or rather to frame", "It helps them understand how flexible people are in the workplace", even if they add that "They already know all this". During these training sessions, what seems to be said collectively is something that has already been mentioned and that the individuals already know. This work on profession or status raises the issue of speech, which must be internalized, and control over one's body (posture, non-verbal language, codes related to the expression of emotions) enabled by playing a role.

In some of the role-playing games observed, the role of the interlocutor played by the trainer or one of the participants may also be discussed, as will be the case for a training session designed for the managers of a large consulting agency. "Personally, I wouldn't have accepted it if the consultant (the manager's subaltern on which the exercise focuses) had started talking shop" (a manager taking part in a learning session). The strict framework of training for a specific role is left aside so that people can consider acceptable or unacceptable solutions, which will be made quite clear; one of the managers taking part in the training will brandish, during one of the role-playing games, a piece of paper for his colleague with the words "bash him!" written on it. During the training session, some counterexamples can be incorporated into the common references: "Don't be a Bertrand" will become the inside joke created by the trainer and then used by one of the groups after a presentation that one of the participants thought was "a failure".

3.2.2. *Deframing/reframing: role as technique*

If we consider business theater, this verbalization and acting out of roles involves, first of all, an external suggestion made by the business theater provider, which consists of a proposal – which some regard as a "caricature" – to represent "cases" or anecdotes collected when this service was requested and while it was being documented (the trainer– actor– scriptwriter generally discusses with two or three contacts provided by the sponsor). This short performance, which will be a prelude to the training session, is called a "sketch" by business theater professionals.

"We're going to play a story inspired from elements that we condense in an extremely short time sequence. It is a group of events that together create a caricature" (an actor-trainer).

From the trainer's perspective, this reduction, similar to a caricature, has two functions:

– it allows one to distance oneself: "We may talk about a certain topic with some perspective, by using this scenario (…). It allows us to avoid approaching a theme too frontally";

– it helps one accept fiction because of how it refers to the actual world: "If we tell them: 'I'm going to play something real', then they'll say, 'But it's fake', whereas if I tell them that it's a caricature, they'll tell me that it seems real" (an actor-trainer).

We may find it paradoxical that access to reality necessitates fiction in this case. Some trainers who employ role-playing games through these real-life scenarios will scrupulously point out at the beginning of the training session that this is "fiction". The fact that this fiction may be pronounced, and even exaggerated, because situations regarded as typical are condensed in a very short period of time, does not prevent it in the least from being perceived as a representation of the "truth".

Another aspect we find interesting consists of the fact that, according to some business theater professionals, "it is necessary to break the codes" (a trainer). The fact that the actors "express emotions that are not the same as those a colleague would express" (an actor-trainer) creates a sense of perspective that allows us to take these codes into consideration. What is staged must simultaneously portray the company by taking up some of its related anecdotes or codes (verbal, dress codes) and be fake enough (especially emotionally) to make the participants in the training session at ease, so that they can make comments, possibly criticisms, and get involved in what happens during the training process. The modeling factors of the role-playing game, which are based on fiction, allow people to express action and behavior norms because of the game's distance from the actual world (in other terms, the gap between map and territory, if we want to use Bateson's image).

We may be led to think that this takes place, through the corrective measures taken, according to processes of interiorization of the constraint. Participants are encouraged to impose rules on their own. "A managing director asked me to stage something that could put people in a spot and make them do what they have to do, so that they behaved in what he considered the right way for his strategy. Maybe we'll come to the same conclusion, but it won't be imposed" (a trainer). By reflecting on the role played, which mirrors his own, an individual is able to formalize "the behaviors that do not work and how to think differently". This formalization is not prescribed by anybody: "My job consists of telling someone: you decide" (a trainer). Another trainer will tell us: "The idea is that it is the participants themselves that should come up with as many things as possible".

The roles proposed by the actors can then be used to work and reflect together, now explicitly, on the role of the professional. When the participants are encouraged to "play" with the actor in "real-life scenarios", their status does not change. "We offer them to be part of a fictional situation while keeping their own roles" (a trainer-actor). They play while embodying their professional role, which is not the case for other games we observed – role-playing games, switching places or hierarchy reversal games in which the other person's job or hierarchical level is involved. In business theater, they may be encouraged to play their own role even in unlikely situations: "We make them play something they'll never see, like 'Well, no, I'm telling you. No, boss'. They can try out certain things in this situation that they couldn't do in their daily life. This allows them to see a lot of things" (actor-trainer).

Actors-trainers, as well as trainers, will often test the participants' ability to react to the unexpected in real-life scenarios. This deliberate use of the unforeseen will be much less common among colleagues grouped together in game contexts during role-playing games. The fictional frame of role-playing games is here strengthened by proposing behaviors that do not comply with the professional codes of predictability and foreseeability inherent to them. In this case, what is fictional is no longer the simulation frame, but also the reference to reality. The situation no longer has to do with a simulation of the actual world, since, in the words of a trainer (actress), individuals are explicitly faced with situations "that they'll never see". We may think that the fact that the real world need no longer be referred to will lead those participating in the training session to work their role no longer merely in relation to their status but *in situ*, in the situation of interaction, which is congruent with the other historical definition of role, namely its interactionist rather than fuctionalist interpretation. According to the interactionist approach, role represents a convergence of social issues of predictability and the actor's ability to take initiative. As we will be told, one of the goals of a training session involving theater consists of "no longer being in control of what commonly takes place in the office of a company" (a trainer). Role, in this case, is no longer part of a mere codified and normative relationship with status, which people should know how to inhabit. An individual is tested in his ability to react or adapt to the unexpected nature of his relationship with the other, while also playing his own role. The problem of status arises again at a later stage, during the debriefing, when we want to find out if, by reacting in a specific way to the unexpected, the participant has stuck to his role or not. The risk of losing control is an integral part of the game and emphasizes in everyday work the part of uncertainty that represents one of its features. Consequently, we can consider the double relationship established between uncertainty and work, and between uncertainty and game, by:

– On the one hand, dealing with uncertainty – professional situations themselves are made of unforeseen events and uncertainty, which getting out of our comfort zone could somehow prepare us for. As one of the trainers will tell us, one of the goals of these training sessions designed for managers consists of making them work with "fluency, flexibility, agility" and to prepare them "to face any kind of situation". We find here the idea (shared by ethologists, like Groos, as well as educationalists like Bettelheim, besides several game theoreticians such as Caillois) that games prepare us for life (Caillois) by increasing our abilities to overcome obstacles and difficulties.

On the other hand, and more specifically, putting our interactions at risk because of uncertainty. Games, in this case, operate by destabilizing everyone's roles in the sense that one of the individuals interacting with us will not stick to his role ("Personally, I wouldn't have accepted that", "he has to do it, still"). By compromising the "predictability" dear to Goffman [GOF 63, GOF 67], smooth interactions are no longer guaranteed and everyone has to save face by adapting his behavior in relation to the situation. There is here a double issue for the participant.

– As a player, he is led to comply with the instructions he has been given at the beginning of the role-playing (pulling a subaltern back into line, convincing him to take charge of an issue he does not want to have anything to do with, etc.). He might lose the game if he does not manage to overcome the obstacles that have been created for him in the verbal sparring among players. Some win and some lose.

– There is another issue: not losing face in front of a group of colleagues, peers, and possible competitors in the company, who are present during the game and will be encouraged to make comments on it, judge the player's performance, as well as evaluate the professional playing, his ability to overcome his adversary or a professional obstacle (i.e. his subalterns in the case considered) encountered as a manager, since, as Goffman reminds us in *Encounters* [GOF 61]: "games give the players an opportunity to exhibit attributes valued in the wider social world, such as dexterity, strength, knowledge, intelligence, courage and self-control" [GOF 61p. 68]. In this case, we may think that managing someone else involves first the ability to control the game and come out a winner.

In the role-playing games considered here (business theater and real-life scenarios during training sessions), status is always referred to: either because it must be performed, as it were, in a fairly codified and normative way (or to correct the actors' performance in this respect), or because we must keep our role/status while compromising in unexpected situations on the basis of our own resources. The debriefing can be seen as the moment in which everyone assesses whether this status has been kept (or not). The new suggestions made by the participants are discussed and role is negotiated.

In this case, we may think that the training frame and its structure (also including the stages of didactic training, which starts with a preliminary phase, a game, a debriefing) makes role lean toward status. The training process prescribes roles and participants will accept what has been suggested to them (with the exception of some individuals who will not play the game, rare cases of people writing e-mails on their computer or reading newspapers on their screen during the training session). The participants master the expected repertoire and will, in the context proposed to them, do what they are expected to do by implementing the "roles to play" or keeping them at all costs. Thus, they comply with the rules of the game and the instructions provided by the game frame.

It is not guaranteed that this will take place in the same way outside the game frame, where statuses may or may not be kept in professional situations. In this respect, more recent meanings of the concept of "role" can be used. Barbara Simpson and Brigid Carroll [CAR 08], referring to Kunda [KUN 92], claim that if many "ready made" roles (as they say) "exist within organizations to communicate how individuals should think, feel and act (…) individuals choose the extent to which they embrace or distance themselves from these roles at different times" [SIM 08, p. 32]. Since the space-time of the game frame is fixed, as well as its expectations, the issue of how people can distance themselves from roles or inhabit them in different ways is usually not considered. Participants are there in order to play their role, to contemplate "what they must or can do" in their position, as they are invited to be trained as managers.

Outside this game context and the rules it involves, namely in actual work situations, we can actually think that status is referred to with more flexibility. It will be mobilized or not by an individual according to how useful it may be to have recourse to it in the situation experienced. We can also consider that a definition of role conceived as "as a vehicle that mediates and negotiates the meanings constructed in relational interactions" [SIM 08, p. 34] does not incorporate what we notice about the game frame. Passing from the territory to the map, i.e. from the model to the simulation, in the case of simulation games, leads us to underline the difference between the transition from a situation of interindividual interaction between a manager and his subaltern to a daily work situation and the semipublic context in which this interaction is simulated. Those taking part in the role-playing game are, in this case, supervised not only by the trainer (and the didactic principles he enounces) but also their peers and colleagues. Staging their actions requires them to immediately refer to the role prescribed and expected from them.

3.3. Asymmetrical reversals: what happens to social relations in the game?

Fiction and role, as well as performance, successfully converge in the practice of reversal days. A reversal day, just like a carnival, relies on a process of "fictionality" [PIE 05, p. 42]. On a workday just like any other (temporal delimitation), we pretend that we are an accountant and that the director of the company is no longer a CEO. At the same time, if all of this is staged, the game remains bound to its game frame: the boss does not sign any contracts for a few hours and his ability to take initiative becomes quite limited in scope. Work, its context, and its social relationships in the reference frame are combined with the artificial aspect of the game (which allows us to meet a short-lived version of a director[6] or car washer). As a game, it requires us to establish certain rules about the distribution of roles (transformation rules: swapping jobs or hierarchical places) and to define its goals, always made clear by directors: becoming aware of what "actual" work is, allowing exchanges between categories of professions, etc. This method is somehow ambivalent, once again carnivalesque, and "allows us to play constantly on two levels while combining what is incredibly serious to what is utterly ridiculous" [PIE 05, p. 43].

Passing from one to the other involves the main marker of fictionality, i.e. dressing up. Embodying someone else's role amounts to wearing his or her uniform and dressing up, which will often elicit laughter.[7] The episodes of "*Patron incognito*" will even involve making up. TV programs based on the principle of a boss turned subaltern dedicate long scenes to the moments in which bosses, while changing their clothes (occasionally taking them off), move to the other end of the hierarchy (and consequently class) and profession. The set of status symbols (suits, brand-name watches, etc.) are filmed in great detail. If we leave aside its media versions, a reversal day is a dressing-up day in the two companies studied. The functional managers or hotel directors working for H, a hotel group, wear maids or waiters' uniforms and the person replacing C's boss for a few hours leaves his piercing at home for a day, shaves, and wears a shirt. From this perspective, the functions of costumes may as well match those of the masks that Bakhtin

6 An exceptional case of "ascending" reversal day, which we have come across only once. The vast majority of them are based – if we consider hierarchical reversal days – on a top-down principle: managers and directors take the place of subalterns.

7 The trunks of an aqua-gym instructor in an episode of "*Vis ma vie*", management dressed up as hotel waiting staff filmed while being laughed at – for internal communications – etc.

discusses when studying festivity in the Middle Ages: "The mask is related to transition, metamorphosis, the violation of natural boundaries, to mockery and familiar nicknames" [BAK 70, p. 49].

3.3.1. *Terms of reversal*

The term used to describe this event, "reversal day", explicitly refers to the subversion of order that characterizes carnival [HEE 71]. A reversal day, just like its counterpart on TV (the series "*Vis ma vie*" (Try My Life) and "*Patron incognito*"), is inherently linked to the question of hierarchical reversal. As the authors who have studied social reversal games (historians, folklorists or ethnologists), namely games involving the temporary reversal of places, point out: "This apparently revolutionary and disruptive phase provides an outlet for the tensions generated by the social hierarchy in place without damaging society in any way. The authors have underlined its function as a safety relief valve and its therapeutic qualities" [PER 67, p. 435].

In the TV versions and, for the most part, in business reversal days, those at the top of the hierarchy will often find themselves carrying out the tasks of "low-staffers"[8]. However, those at the bottom of the work ladder, with the exception of one case and under certain conditions, can never have access to high positions. Similarly to a carnival, we can see quite immediately that the reversal has nothing systematic about it. Following the example of traditional [LAR 86] or Creole carnival [MAU 10], it is possible for men to dress up as women (which is also an element that elicits laughter). Similarly, in the advertising agency C, one of the employees decides to go to work wearing a skirt, high heels and a generously low-cut dress (sic) to take the role of receptionist. The individuals we observed, regardless of gender, also feminize men who are doing jobs done by women. Colleagues will joke about two managers replacing some accountants for a day: "Oh, they're cute". Women do not dress up as men. They wear neither wigs nor suits. At C, we can identify three types of dressing up: gender-related, i.e. men dressing up as women, hierarchical, i.e. a subaltern dressing up as a manager (shirt, elegant shoes, close shaving), or profession-related, namely from salesperson to graphic designer or developer, for instance (or from functional to executive at H). As we have already underlined, the fact that women do not dress up as men, or bosses as employees – apart from wearing a uniform

8 And not those carried out by middle management, which is quite interesting.

when it is required – recalls the carnivalesque practice of power subversion: a male "pretty" manager is feminized into an accountant; a king for a day is elected among the people and his role will consist of imitating power without having to actually exert it. Those in power become subalterns – a CEO becomes an accountant – and the subaltern turned king for a few hours irons his shirt exceptionally well…We can try to imagine what it would be like to mock women dressed up as men or subalterns turned managers, which would most likely be less acceptable or unifying. A reversal day, like carnival, is not exempted from norms and decorum. At C, as well as H, dressing-up games involve the typical attributes of a professional, such as the uniform of a maid or the relaxed look of a graphic designer. In this respect, dressing-up cannot be exclusively seen in relation to mockery or social relationships of class. It also highlights the identity function of carnival [PIE 88, PIE 05]. The attributes specific to typical professions, while being overemphasized, are pointed out as part of people's identity on this day. Finally, in games involving power subversion, the object of mockery constitutes something that it is possible to laugh about, in light of its related social privileges.

3.3.2. *Limits of reversal*

Reversal, like carnival, is well "outlined only in the most immediate and superficial symbols" [BER 94, p. 32]. As Michel Agier [AGI 00] suggests in his observations about carnivalesque practices, carnival involves distortion and detachment from reality rather than reversal. In this respect, he agrees with Barbara A. Babcock, who underlines how: "Symbolic inversions are not simply logical reciprocals; if you consider them as such you tend to neglect both the transformations which occur with such inversions as well as the comic dimension of many such reciprocal forms" [BAB 75, p. 157]. Thus, we could regard these gameful-carnivalesque forms as disconnecting institutional social relationships of domination. This disconnection is combined with the replacement of work relationships, presented as converted into games, with a festive and occasionally comical event. This takes place while, as Marie-Anne Dujarier underlines, "work is often presented as the exact opposite of games, which are the activities that most represent 'off-work time', leisure, and 'free time'. Besides, they are forbidden and even punishable during working hours" [DUJ 12, p. 90]. Hierarchical differences are abolished by the game because of the reversal processes and the fact that employees are on the same level as their managers ("*Vis ma vie* (Try My Life) – *Mon patron à ma place*" where employees see their boss come to work with them on the assembly-line or in the kitchen). In the latter case, people laugh with their boss about his or her failure, while this laughter cannot be shared when the director takes someone else's place in his absence, which is what happens during reversal days and, in a certain way, in the episodes of "*Patron incognito*". During this event, held at H, it is managers who laugh a lot, and the human resources director underlines the "terrific atmosphere" that people breathed at this event which ended up being a friendly meet-up. This management initiative borrows the principle of "not for real" from games: "A player 'pretends' that objects and roles are true, while also being joyously relieved by the knowledge that it is not so" [DUJ 12, p. 93]. The "as if" of mimicry is in this case total: people pretend to play as well as work, even if they are taking someone else's place.

However, this process differs from carnival (and festive) events in one aspect: employees do not necessarily take part of their own accord. Reversal days, under the guise of games, actually involve a kind of work that includes the benefits associated with the "pleasure" of playing. Both the hotel manager of H and the employees working for C have obligations. Only the head office management of H are free to choose if they want to participate or not and point out their preferences. At C, roles are assigned at random, which is another defining feature of the game. However, people make sure to remove certain names on the sly when the post of director is randomly assigned.

Another difference, in this case from games, has to do with the criteria of lack of consequences and unproductivity [CAI 67]. We have seen how, in several theories about games, a game is regarded as unproductive and with no consequences on the real world, which is an issue that has then been discussed. In these theories, games are played "for nothing", at least in terms of material effects on reality (for example the actual consequences of buying decisions when playing monopoly, since gambling is no longer considered relevant to games in this respect), unlike work. Even if it does not follow a logic based on short-term profitability (since there is no one replacing the players at work on that day) and may be considered as a day of "losses" for a business, a reversal day involves explicit management expectations, which are explained to the whole staff before it takes place. The goals stated have to do with intercomprehension and teambuilding. A more implicit issue, which concerns facts rather than the management's speeches, may consist of how games allow people to get involved in someone else's work, when it is management that "play" employees. The managers of C as well as H expect to learn something from the game. Just like in carnival, serious and comical elements are combined in the interrelation between games and work.

In this respect, a reversal day is paradoxical: it employs the concepts of festivity and authentic relationships in a work environment allowed by the topsy-turviness that characterizes this specific day (being close, understanding the other as well as possible) and attempts to affect social bonds. However, it makes it look like a professional constraint, which is supposed to give it a meaning in a work context where playing just to play is not tolerated. In this management process, festivity is combined with constraint and business methods are at the center of the exchanges that are being encouraged. The personal aspects involved become institutional. Unlike other internal communications events, in this case the goal of reversal days consists of consolidating work links, rather than sociability and links at work, which are strengthened, for example, during meet-ups or parties. The links involved have to do with profession or hierarchy: relationships between a graphic design and a salesperson, or between a manager and a subaltern.

3.3.3. *Taking another's place up to what point?*

As they are being used, reversal days are still based on a hierarchical principle: in the company H, managers are the only ones who can take over the role of their subalterns[9]; in the company C, only managers can refuse to be replaced, they are not briefed, they can lock their files, and they are allowed to get involved in someone else's work. For the CEO of the company C, this day represents an occasion to identify who plays along and who does not, and consequently to evaluate his coworkers: "This allows us to see how people use their new function and also how they deal with certain kinds of management" (CEO).

If the low-staffers working for H who do not get involved in this activity are actually passive since they are left out, we can notice that the employees working for C are also passive, to a certain extent, in their 1-day workplace. None of them runs any risks or takes initiative in relation to someone else's job. No one decides anything that could somehow affect the job of the

9 In an outstandingly devaluing relationship where only managers can do someone else's work but not vice versa [JEA 08].

person replaced. At the end of the day, each person will be judged by the individual replaced according to the final result and everyone will admit, in kind tones, how difficult the tasks of others have been.

Albert Piette refers to the "paradoxical implication" inherent to carnival, i.e. what happens during a reversal day in which we can find a "double constraint" that drives those who are acting to, as we have seen, "avoid going through with each type of behavior" [PIE 05, p. 45]. If we consider reversal days, it will be the subalterns, rather than the directors, who enact the role assigned partially. If the "substitute" of the boss of the advertising agency takes any decisions, he will do so as a joke and he will not argue in favor of them during the meeting at the end of the day on the basis of his entitlement. If he had done that, he would have pushed the principle of the "as if" past the boundaries of the game by producing certain effects on the actual world. Therefore, he plays a game where he takes decisions "for nothing". His colleagues at the agency will also make sure not to encroach on the activities or professional domain of their coworkers and will merely follow the instructions provided. On the other hand, as we have said, the two directors of C take action in relation to the posts they have inherited and change, on their computers and even in their files, the methods used. They are also the only ones not to be "briefed" beforehand, as the rule that they have set requires, and will be given no instructions by the two accountants whom they are replacing that day: "She didn't dare" will say the CEO smiling, when he is asked the question.

In the morning, the discussions between "briefers" and "briefed", before everyone takes over someone else's role, are mainly focused on recognizing the difficulties involved in the position that people will have to take over and their own inability to fulfill it. The comments made at the end of the day will highlight everyone's inability to carry out as efficiently what the person replaced does, subject to the approval of general management. Thus, during a debriefing meeting, a developer will say that, according to the instruction he has been given, he has "chased up three clients" by e-mail. "It would've been worse over the phone", he will say amused. After his remark, the CEO will say to everyone present: "We have lost three clients", and everybody will burst out laughing. The only one saying that "it was easy" (the legal intern) will be mocked by the boss while everyone approves.

Professions that are regarded as menial, such as chambermaids, will also be appreciated because of the event. The fact that the "functional" managers working in the head office of H spend a day cleaning hotel rooms represents a token of gratitude for "menial" professions. Thus, thc human resources director will tell us that: "It's really paying tribute to the small hands that are invaluable for our professions. It was really about putting a low-staffer back to the center of the business project, this was our goal, and it's been achieved. (…) I regularly meet up with low-staffers and they find it terrific, it's great! (…) We have diminished our low staffers' turn over, which really increases their involvement and motivation". Except that, while "paying tribute to the small hands", as it was phrased, what comes across is the exact opposite. This interpretation is based on the assessment of significance (the small hands), rather than the technical skills or expertise specific to each profession. Going down the hierarchy could correspond to a form of condescension since, if in group H everyone can "stoop" and do the cleaning, no one would be able to "go up" and be a manager. This happens because the reversal day is, within the group, a one-way process.

3.3.4. *Changing place, changing view?*

At H, what managers often talk about at the beginning of the reversal day is the way a profession can be experienced. The (short) interaction with some employees, and consequently the experience of the profession, pertain to this topic. "Our purchasing director, who was a maid, went and spent two hours with a maid during work, when she had some time, to work together and see what that was like. (…) Everyone received two hours of training twice before the event". Similarly, the CEO of the advertising agency says that the event allows us to "become aware" of other work constraints besides our own and change our perspective. "A reversal day allows us to get some more perspective in relation to how we see others: which is "cool" like being out driving for a salesman. Yeah, but to what extent? If at 7 PM you're still driving and you're far from home, it's less cool. They show to each other what each other's job is like: an accountant with a salesman, a graphic designer, etc.".

The time span itself of reversal days, just like the one of the experience made by directors in the reality TV programs studied, introduces bias in this logic of experience through role-playing games. It seems difficult to expect people to be able to experience in a day the actual work of employees, unless we considerably reduce the elements in which their activity consists. Moreover, spending a few hours doing someone else's job seems fairly different from the actual work conditions and experience, namely the effects on mind and body of routine, repetitiveness, and the patterns of the actions performed, without forgetting the consequences related to the application of quality or safety "productivity norms" established by the company.

This bias is increased by the leeway given to those participating in the event: "All the managers that used the game have been able to choose if they preferred making their experience cleaning or in a kitchen. Then, they organized their work schedule" (H). Besides, the tasks carried out, in this specific instance, are overmanned: "They needed a lot of people to do it, more than the usual coworkers, it was quite different from the productivity norms". The nature of the work experience undertaken this way can then be called into question.

The experience provided by the reversal day taking place at H represents first of all a solution to the lack of field knowledge of the functional managers, who do not know much about the tasks carried out by the operational staff: "But let's think about one of our professions, which is actually the toughest one, maids. I think that among our salespeople, if some of them found it a bit contemptible to clean rooms, when they've done that for eight hours, I can tell you there was a lot of respect and admiration for these women who do that on a daily basis" (human resources director). From reversal days to reality TV programs, making people more aware of the jobs of others is one of the goals constantly pointed out. However, bringing the different actors of the business closer represents a second issue: "For me, one of the good things was that operational staff and functional managers got closer (…) since not everyone comes from a hotel business background; there are people who had other jobs". Finally, this operation allows managers who have reached high positions to get to know the field better and the gap created by the hierarchy to be bridged: "Then we listened to what our chiefs of staff told us and we realized that our hotel directors, who had been chiefs of staff at a certain stage of their career, but sometimes had maybe for a bit, can I actually put it this way, not completely lost touch with the reality of everyday work, but anyway they had a perspective that was no longer adequate…" (human resources director). In this sense, the experience results from examplifications. It may be the experience made by employees who see the managers of their company required to carry out their tasks and getting closer with them. The issue of performance is omnipresent for reversal days, which turn out to be large internal and external communications operations, all the more so when they take place and are shown in mainstream reality TV programs.

Representations and the attempt to modify them by staging them are also at the center of this operation: changing the way functionals or hotel directors see operational staff, of course, as well as hopefully altering the way subalterns view their management: "Well, they (the maids, *ndr*) have seen me. Otherwise, they would've only received my feedback but now they've seen me live on TV while I was cleaning (laughter) and that's been an extra booster for them" (hotel director). At C, the operation also involves the issue of modifying the way the boss is seen (who thinks that he is regarded as a "hatchet man", as he will say) as well as what he does and the difficulties associated with it. Replaced by one of his employees, he will tell us that he is happy to see that his coworker is finding out how difficult his job of director is. An expected recognition of the CEO's job is implicit. The individual elected as a director for a day, complying with the behaviors required, will make a public speech during the collective "debriefing" that denies the comfort and privileges associated with the position of director: "I thought I'd have fun…give a pay raise to everybody, buy a swimming pool…but, actually, no. I was a CEO today" (L., his substitute).

3.3.5. *Carnival and order*

A manager, after carrying out himself the tasks considered as the toughest or most menial with a smile on his face, gains greater recognition and can demand the same behavior that he adopted during this day, which is filmed, diffused in hotels, or broadcast on TV channels. The smiling attitude displayed by managers, in order to be exemplary, is also a vehicle for the demand of "social skills" specific to waiting activities.

The hierarchical reversal of roles consequently allows us to deprive coworkers of their chances of withdrawing. The process reduces them in part to silence, in relation to any potential management misunderstanding about the actual reality of their work.

Unlike H, where there is no need to involve a coworker to represent his or her job after the two previous hours of training, at C, involving a direct job swap to "become aware" avoids the mediation of management concerning handling conflicts. In both cases, we can see the assumption that understanding the other's job is necessarily a matter of empathy, but in a way through substitution. Taking someone else's place can in this case be taken in its literal (taking his or her place) and non-figurative (adopting his or her point of view) meaning. For the specific case of hierarchical[10] reversal days, the assumption is that any experience corresponds to a principle of standardization of work specific to a Taylorian approach to a so-called execution type of work. The fact that, in hierarchical reversal days, it is the professions at the bottom of the hierarchy, rather than the middle management ones, that are involved seems to emphasize the distinction, made by management, between those who are replaceable and in which function (and those who are not). While one of the principles on which the operation is based is intercomprehension and the approach to someone else's experience (the title of the program "*Vis ma vie*" (Try My Life) will be regularly mentioned in companies), the approaches centered on the concept of reversal day seem to lead to a limited interaction with the multitude of analyses and individual perception in the workplace.

From a critical perspective, we could claim that this operation allows the hierarchical structure in place to be forcefully re-established. These structures may be authoritarian. This is the case for bosses getting directly involved in someone else's work without finding out about his or her experience or asking advice and, maybe, for the imposition of the management experience specific to someone else's profession. They may also involve the management's promotion of examples (self-promotion) and recognition of everyone's (distinct) place.

Reversal days involve the "normalization of social relationships of production" channeled by management [FLA 12] as well as the trivialization of the social value assigned to certain hierarchies and professions. A reversal day as performance, rather than celebration, aims to act on representations: of directors as not so bossy or indifferent, but humble, of modest but respected maids, and of an organization where it is enough to see things in a different way to relieve tensions.

These reversal games simultaneously represent what it means to be a good employee and what it means to be a good boss. A good boss knows

10 Since, as we have seen previously, the possibility of experiencing a colleague's job is denied in reversal days that encourage job swaps between different professions: too tough, etc.

how to remain close with his employees, how to be spontaneous, and would not hesitate to join in the work and leave the attributes and privileges associated with his position of power. Dealing well with the obstacles encountered while doing something he does not know about (and faced with his social downgrading), the boss playing the employee represents an example of what it means to be a good employee by being positive and proactive while carrying out tasks perceived as lowly or menial. Conversely, the employee embodied by the boss (as well as by a colleague during professional reversal days) is quite pathetic: for example, he is hindered by his incompetence or lack of knowledge in many ways and turns out to be rather inefficient. The principle of the necessity of everyone's place has been re-established.

Reversal games reproduce three aspects of carnival:

– they represent a brief interlude during which the social order is disrupted, which leads ritually to its re-establishment and the re-assertion of the principles of order [BAK 70]. In this game frame, comedic reversal, the "grotesque reduction" of power [BAK 70], and its implementation (deciding when the process takes place, getting involved in someone else's work, for instance) co-exist;

– they offer a blend of the hierarchy's antagonistic dimension and its emancipation, comedy and seriousness, work and game;

– they follow a principle of involvement based on disguise or masks, which allows people to inhabit the role of the person embodied: it is a principle of immersion.

Even if we have noticed the carnivalesque laughter associated with the grotesque reduction mentioned by Bakhtin, the latter encourages us to distinguish between the domain of our observations and the popular festivals linked to carnival. The fact that they are organized top-down and that not everybody takes part in them leads us to reintroduce the difference between game and play, structure and behavior: "The feast had always an essential, meaningful philosophical content. No "exercise" related to the management and perfectioning of the process of collective work, no "game at work", no rest period or breathing spell can be rendered festive per se" [BAK 70, p. 17].

However, it remains interesting to observe how management, making use of these reversal games, is empowered, recycles, and comes to a better understanding of the traditional forms – which are fundamentally popular – of the notion of world, order and disorder. Bakhtin's statements lead us therefore to question what, outside a festive context, games turn into during the gamification process, i.e. the functions induced by applying the structure of games to work environments.

3.4. The game as an operating structure and the performativity of the game

In his work called *The Ambiguity of Play*, published in 1997, Brian Sutton-Smith [SUT 97], a researcher in learning sciences, recognized that Western research has presented games historically in relation to seven dominant discourses (rhetorics). Drawing from those who have written about games (let us mention the most famous ones: Groos, Piaget, Winnicott, Vygotsky, Huizinga, Abt, Turner, Babcock, MacAloon, Schechener, Goffmann, Bakhtin, Bateson, as well as Pascal and Heidegger), he identifies seven categories used to conceive games[11]: games of development/ adaptation or progress (in relation to childhood); games linked to chance or fateful coincidence (the most ancient discourse according to Sutton-Smith)[12]; games as power (leadership in competition or conflict as well as the "force, skills, and leadership" used during the game); games as collective and identity-related celebration; games as imaginary, i.e. a creative potential driving change; games and self in relation to development, as well as the performance of the game (excitement as well as flow) and, finally, games as linked to frivolity.

We could say that these discourses follow two types of logic:

– a logic of definition in relation to how games may be associated with chance, identity or frivolity. These discourses make games a vehicle for these elements;

– a logic of action. This is the case for discourses of progress/adaptation, empowerment, change in an imaginary dimension and development of the self. These discourses give games a dynamic dimension in relation to the other, the self, or the context. Sutton-Smith mentioned that, in his opinion,

11 We mentioned them on page 43 according to Gilles Brougère's reinterpretation of them.

12 Sutton-Smith points out here the paradoxical relationship with the more contemporary notion of games as a product of our free choice.

the modern discourse on games focuses on the concepts of development, self and imagination.

Deterding [DET 14], drawing from the list of game-related discourses made by Brian Sutton-Smith, puts forward a new classification of them, which he considers as "practical tools" as well as Weberian "ideal types" [DET 14, p. 34]. These discourses can be found in the professional research and practices related to gamification (designers as well as consultants). The 14 categories he identifies are distributed based on Sutton-Smith's classification, except for the category of chance, which is absent from his list:

– the category of progress includes discourses about;

- feedback, which is regarded in terms of transparency and meritocracy. Deterding points out the similarity between this discourse and the world of work: "It fits nicely with mental models and practices dominant in enterprises: games are seen as business dashwords and incentive programs, only somehow more and better. The societal role of games is thus framed fully in luminal terms: game design can allow managers and regulators to arrange for more competitive and productive employees (…)" [DET 14, p. 39],

- nudging (encouragement), which has to do with the goal of persuading, i.e. the ability to encourage individual choices so as to orient them in a desired direction (in this respect, he refers to "choice architecture"),

- hedonic development: resulting from the "necessary autonomy" linked to games, which favor flow (Csikszentamihalyi, McGonigal) and personal well-being,

- sociotechnical systems that favor learning,

– the category of power;

- exploitation, namely gamification as a product of capitalism as a response to the rhetoric of feedback and nudging,

- "status", i.e. the notion that the recompense everyone strives for consists of obtaining a higher status and better social recognition,

– the category of identity;

- "collective performance" regarded as the collaboration and sharing of experience, which gives rise to collective, and even community (for example communities of players), experiences and re-assertions,

– the category of the imaginary;

- immersive performance combining mostly role-playing games and augmented reality,

- the systems that Deterding calls "expressive", i.e. "meaning-making" media [DET 14, p. 45]. These systems, according to him, lead systems of rules to interact with people, which creates the category of "persuasive games" [BOG 14]. Deterding regards them as simultaneously opening up the possibility of comparison (and becoming aware of them, since the rules are supposed to be made clear by the game, which is paradoxical when we consider the aforementioned practices of "choice architecture"),

- cultural form, which amounts to saying that games, as an expressive and aesthetic form, are part of the history of these forms and spread the meanings that are specific to their cultural domain,

- playfulness: The "play" element supplied by the game, which constitutes its foundations. Deterding calls this type of "play" "industrial", since it feeds into the works of the "ludic design" industry,

– the category of self;

- eudaimonic or the well-being linked to the development of the self. In this case, well-being represents an accomplishment linked to knowledge and self-knowledge, as well as the mastery of personal skills and potential, which is what games make possible,

– the category of frivolity;

- the pleasure associated with the game experience,

- the "free" playfulness (in line with Caillois' concept of *paidia*) that manifests itself in the creativity and emancipation from the rules that it allows.

Out of the 14 types of rhetoric that Deterding identifies, eight have to do with discourses on progress and the imaginary, which seem to prevail over those, for example, on identity, self or frivolity. Strangely, chance does not figure that much in the discourses on gamification, which seems to back up our hypothesis about the centrality of a directed performance: in these games there is no place for letting loose, games do not take control (as is the case for games of chance), but are under control. Most of the categories identified by Deterding have to do with the idea of performance, individual (and collective) development and productivity in terms of pleasure, self-realization or self-knowledge, and increased personal skills as well as knowledge. It is this performance that accounts for the sociotechnical equipment used to accompany, motivate and evaluate an individual. The gamification approach recycles the Marxist concept concerning the possibility of going beyond a kind of work limited by technique (and by itself). The notion of empowerment linked to gamification regards the pleasure or development dynamics supposed to be inherent to games as central to the process of emancipation, which corresponds quite precisely to the new capitalist mindset that praises self-development [BOL 99]. To a minor extent, we can find the concept of pleasure (not including the goals of well-being and individual development) and leadership, which distance these discourses on games from the play dimensions and the political, aesthetic or social criticism of their existence and application.

A traditional functionalist approach to games dominates their related discourses and justifies their use in non-game contexts (the non-game contexts associated with gamification). It is this set of rhetorics that sees them as services or tools available to management.

If we have to analyze the logic of action underlying the games that we have studied, namely role-playing and simulation games, we will take up the speeches made during the training session or games observed. According to them, games, taking place as structures (performative dimension of the game), allow us:

– to create experiences which, as we have seen, are at the center of the use of role-playing and simulation games. Moreover, unlike the "actual world" or reference frame, a game and the frame it offers allow us to test our own role, as well as someone else's, without any consequences;

– to train train our responsiveness and flexibility, and to explore more. To develop or widen our range of methods and skills when we are faced with the unexpected;

– to be efficient in relation to our self-realization: when the confrontation involved in the immersion increases our own efficiency and personal development;

– to speak up, which is allowed by the fictional frame, and its consequence, i.e. emotion;

– to favor learning processes.

However, gamification involving role-playing and simulation games allows us, more implicitly, to reorganize the professional world through:

– the adoption of rules and norms: a situation in which playing the game amounts to accepting the order of the game.

3.4.1. *An experience without consequences?*

One of the main ways in which games are supposed to function, which is mentioned in all game theories as well as in the speeches made by the trainers and managers who use games, consists of providing "experience without consequences". In relation to the games we have observed, this function can be broken up into two elements: experiencing, in the game, either the role and profession of someone else or unexpected situations, and testing the potential linked to one's own role or relationship management techniques. These experiences and experiments are at the center of the practices of role-playing and simulation games. The "not for real" principle of the game allows us to train, to vary our strategies, to experiment, all things considered, without any consequences in "real life" or, in any case, in the context of an immediate professional action. The issue of the lack of consequences, which, as we have seen, is regarded as one of the components of games (as fictional frames) in the theories about them, recalls, in terms of design, the fact that "we play when we are safe" (Tim Brown, Ted talks)[13]. It agrees with the justification of simulation processes for risky activities such as medical practice, the army or aviation. This experience is somehow deferred by the artifact represented by the game frame.

In relation to the games we are studying, designed for managers in work environments, the simulation device represents a social device. No equipment is tested or tried, it is relationships and interactions that are at the

13 http://www.ted.com/talks/tim_brown_on_creativity_and_play.

center of the simulation. The use of games involves the notion that what takes place in the actual world will be reproduced in them: "(...) a game is a kind of play upon life in general, it induces, in a restricted and well-defined context, the same kind of motivations and behavior that occur in the broader contexts of life where we play for keeps" [COL 68, p. 20]. We actually distinguish between the consequences inherent to the game, which, in light of its fictional dimension, involves no professional actions (for example making a sale, recruitment, or evaluating a coworker are actions performed in the game but not in the real world)[14], and consequences external to the game. We could question the total "inconsequence" of the latter, taking into account that it takes place in a professional context. What does "not for real" mean, in terms of consequences, when we are dealing with immediate colleagues or with our bosses? Which aspects of the game are kept in cooperative or hierarchical relationships when we have been "bad" or when we have lost? When have the interactions been challenging? For the games we have studied, it is difficult to conceive games without consequences since game aspects are not the only ones involved (like in several game situations, however, played in a family or among friends) and the context of the players' relationships is not merely that of cooperation or entertainment. By superposing game and work frames, it is questionable to conceive them as watertight, since the game is a tool within a work process. Issues linked to the projects we are in charge of, issues of power, and issues of image are, among other, elements that detract from the notion of inconsequence related to a game used in a professional environment. A few trainers are aware of this and advise their clients, for examples, not to let bosses and subalterns play certain games together. Another form of their vigilance can be seen in how they take action when relationships become tense or in how they manage to rephrase and tone down some of the participants' statements.

If the problem concerning the lack of consequences – which we cannot solve in this work – arises, the precept of experience and experiment is maintained. We have been able to see how the rhetoric of "immersion" was regarded as legitimate in terms of the access to knowledge and training that it enables. The mechanisms of simplification (or complexification, if we consider those cases where trainers create unlikely situations or interactions) induced by modeling a context or "case" are rarely analyzed. For example, these types of modeling omit, as we have seen, aspects of the history of the

14 An aspect that calls into question the nature of the game in the role-playing games used during recruitment sessions, which we do not take into consideration in our work.

company or the individuals themselves, which affects or accounts for behaviors, stances, and methods. Some interactions, which can be explained in terms of this history, become less comprehensible in a recreated frame, where they are watched by an audience and trainers. Abt [ABT 68] underlines the tensions or, as he calls them, "distortions", present in every simulation game, between realism and simplification, concentration (of elements) and comprehension (the instructive goal of the training sessions) and, finally, between emotional involvement (required) and analytical composure (also required). All of this raises the problem of the experience made and where it fits. Is it a type of experience of actual work through the "not for real"? Is it a kind of access to the system of norms inherent to the world of work through the figurative dimension of the game?

3.4.2. *Training for reflexivity, flexibility and exploration*

Besides experience and testing by immersion, another topic explicitly linked to training sessions involving role-playing games (this does not apply to reversal days) is the training associated with the practices of one's profession. Therefore, managers are trained to cope with psychosocial risks, to negotiate, to deal with difficult individuals, about security and issues related to discrimination, as well as change management. It may seem strange that managers who have been carrying out management tasks every day for 20 years can be offered training. However, training is supposed to not only provide them with new ways of seeing things by observing what their colleagues belonging to the same rank do, but also to face them with the unexpected and demand a certain responsiveness in relation to their actions. The fact that these training sessions take place in groups and that, for those using business theater, some short sketches can be created and then staged at the end of the day brings about a performance of the self[15] and literally displays to the others what an individual has done (at least in the game frame). The debriefing, as we have seen, finalizes everyone's interventions and is used to discuss which methods are right and which are less appropriate. The goal consists of "getting some perspective": "we can step away from the urgency of these permanent requests to see what we no longer see (…) we can also identify certain feelings that we couldn't quite put our finger on" (trainer).

15 It is sometimes stress inducing. The participants hesitate before "having a go" and the role of trainer consists in encouraging them and even showing them.

One of the elements that characterizes games and is emphasized once again by theoreticians could eventually represent a dominant factor in relation to its functions and its importance for management in terms of training managers: the dimension of uncertainty. Unlike the issue concerning winners and losers, inherent to competitive games, those linked to the role-playing games that create tension (which visibly keep the participants interested) have to do with how the events unfold. How will the trainer-actor deal with the participant's suggestion? What is suggested by certain sketches staged in front of management at the end of a training day involving theater? Learning could here mean learning about the others and certain situations, knowing how to get ready for them, react, and adapt to them. It could also mean knowing how to adapt to what the game makes possible: playing "extreme" behaviors or situations, reacting adequately to the "blows" given by certain trainers and, more rarely, by the participants themselves (bursting out crying, threatening to commit suicide, refusing to do something, leaving, etc.). As we have seen in the section dedicated to role, a game involves the possibility of testing something as much as possible. What has been widened is the range of possibilities.

Many of the techniques used during these games rely on this need for flexibility. One of the trainers-actors explains to us that, as a warm-up, he often uses the "goalkeeper game". In this game, an individual stands between two chairs like a goalkeeper. The others in front of him have to "answer him back". The goalkeeper has to reply to what he has been answered. If he does not manage to do it, the others score a goal: "He must be on the same level and reply along the same lines. Someone else will answer back something else that's completely different, another emotion. The aim of the game is 'I accept everything that's been suggested to me and I reply to this. They deal with objections, reasons, facility, agility. They accept 'every kind of situation'" (trainer-actor).

Another aspect of training consists of exploring (encouraging to explore) some themes that in principle have nothing to do with the trainees. As one of the trainers will tell us: "Training sessions are often imposed, otherwise there's certain topics they wouldn't talk about. Sometimes it's some topics they think have nothing to do with their activity, which actually affect them directly. For example, values, it's something that everyone finds crappy! The same goes for disabilities. We get to look for them because it'll be fun and theater sticks in people's minds. It's always effective to talk about interpersonal relationships. We're straight away at the heart of the matter" (a trainer-actor).

Just as designers talk about augmented reality through the superposition of two-dimensional or three-dimensional images onto our perceptions, establishing a game context allows us to increase our abilities to understand, comprehend and react to relationships, perceptions and interactions. The fictional frame inherent to a game (its relationship with what is unexpected, unlikely, or absent from our concerns, "outside the frame"), as well as the performative aspect, allows us to discover "augmented abilities" in terms of self-analysis and flexibility in our interactions.

3.4.3. *Performance linked to self-realization*

The most central aspect of certain simulation processes is the engaging dimension of the game. Games require players to take action. They become stakeholders, elements of the game and they affect how it develops as well as the resulting teachings. Thus, they introduce variability (which is relative if we consider the expectations in terms of status seen in the section dedicated to role) and sometimes disrupt the "case"-based logic used in these training sessions. Getting involved in the game is what several managers and trainers answer to justify why they choose role-playing games, Kapla blocks games, as well as business theater: players are engaged in the game and it is this involvement that makes learning an easier task. We find here the arguments supported by educators. The passivity of the body, and possibly the mind, when confronted with a didactic method are in direct opposition with the action of the game and its dynamic components, such as Brougère's principle of decision (in a game people make choices) and the dimension of uncertainty[16] responsible for the tense nature of games. Performance is related to players and, as we have seen, the role of manager played during the training session. The same can be said about reversal days and their TV adaptations. In these situations, the manager or director is an "actor" in the literal sense of the word: an actor in a performance given to someone else as

16 Besides the unfolding of the game and taking into account the reactions of the person interacting with us, another element of uncertainty may consist of the assessment of our own performance.

well as an actor of his own role faced with the game situation (and the requirements it entails: playing a cleaning lady, reacting to a burn-out).

In these games, there is a recurrent element of comedy (which will be emphasized by music in reality TV programs) or laughter: the comedic aspect associated with the powerful getting to grips with cleaning tools, the laughter linked to surprise or the actors' game during the training sessions. Trainers regularly talk about "having a good time" during these sessions: "It's designed so that we can have fun", "You have to be able to laugh about problems and see them in a positive light".

Several authors, like Gilles Brougère, Jane McGonigal, Sebastian Deterding, etc., refer to Mihaly Csíkszentmihályi and his concept of "flow" or optimal experience. By reading these theories, it is possible to get a better grasp of why present-day management are so interested in games, since it is thought that the development of the individual, engagement with work and the performance of the company are interconnected. According to Csíkszentmihályi, quoted by Brougère, what the optimal experience that can be found in a game provides is: "more efficiency, creativity, development of abilities, self-esteem and stress reduction. In short, it contributes to personal growth, it is delightful, and it betters life quality" [BRO 05, p. 100]. Achieving flow is guaranteed by: the challenge represented by the task that has to be carried out, which remains still feasible, concentration, a clear goal, a system of immediate feedback, marked involvement and lack of distraction, and control over one's actions. "Personal concerns disappear but, paradoxically, the sense of self reemerges strengthened from the optimal experience" and the exact perception of time vanishes [BRO 05, p. 100].

We can note here the connection, allowed by the game, between dedication to the task and self-realization. Games allow efficiency and well-being to be compatible, since they provide players with the knowledge of themselves required to remain in the right state of flow. Challenge, concentration, target, action and feedback, engagement, lack of distraction, control, selfless in relation to the activity assumed to reinforce/develop the individual, and losing track of time, which are all considered as features of games, outline an "ideal at work", if we want to use Marie-Anne Dujarier's expression [DUJ 06].

Gamification represents a nearly ideal perspective about the possibilities offered by games to make people approach their jobs again without any estrangement. One of the questions on which McGonigal focuses in the beginning of her work *Reality is broken* is this apparent paradoxical aspect of games: "We don't normally think of games as hard work. After all, we play games, and we've been taught to think of play as the very opposite of work" [MCG 11, p. 28].

Games and work exhibit common traits but their relationship with constraints could not be any more different: effort, concentration, overcoming obstacles, time allotted. McGonigal, drawing from Suits [SUI 05], underlines the positive relationship between players, who are highly involved in their game, and this same constraint: "Playing a game is the voluntary attempt to overcome unnecessary obstacles" [SUI 05, pp. 54– 55].

McGonigal conceives games as power to transform. Turning work into a game allows us to combine fulfillment with efficiency: "The solution seemed obvious to Csíkszentmihályi: create more happiness by structuring real work like game work" [MCG 05, p. 36]. Burn-out and addiction are risks mentioned (if somewhat quickly) by McGonigal in this relationship between games and flow, even if the central principle of her works relies on the assumption that our happiness depends on ourselves only and that achieving this happiness involves "working hard on activities that produce their own reward", without many constraints: "And the harder we work to experience intrinsic rewards, the stronger our internal happiness-making capabilities become" [MCG 11, p. 49].

This logic is based on the combination of the autotelic dimension of games with the individual's self-realization. The game mechanism entails the well-being mechanism. Csíkszentmihályi hypothesizes that, specifically, autotelic activities open up for the individual a "new reality", "favor a high degree of efficiency" and "make the self more complex and drive its growth" [BRO 05, p. 100]. We may think that from this highly individualized perspective, other people become an element of the game: either an obstacle to overcome in the game, or more positive game elements like allies. According to McGonigal, it is a matter of social connections in a common process.

As for the autotelic principle of the game, what does it bring about when it is applied to work? A worker whose only gratification consists of even more work? McGonigal seems to agree when she mentions, as an example, the game World of Warcraft, explaining that the real recompense of the game is to be rewarded with "new work opportunities" [MCG 11]. It is all a matter of proportion, which represents the goal itself of its conception: we have to conceive challenges that are sufficiently tough to keep people interested and moderate enough to produce no anxiety or lead people to give up.

One of the properties of games, according to trainers, consists of keeping people interested "since slides are useless". Concentration on the task at hand in front of someone else (a challenge), together with a context that encourages humor and fights boredom, is an element of these game dynamics (gamification) praised by their developers.

3.4.4. *The power to speak granted by the fictional frame*

Certain elements pointed out by theoreticians – freedom according to Caillois and frivolity for Brougère – have to do with the issue of the lack of consequences in relation to what takes place within the game frame. Since it is "not for real", we could do, by virtue of experience – here, in this specific instance, what we would not do in other circumstances. For example, people will sing a little song in groups of four, at the end of a training day at the hospital, to make the administration manager aware, in a polite way, of the tension resulting from the work conditions she has imposed. In the agency C, the CEO for a day will pretend that Whitmonday, a workday, might turn into a day off again by announcing the good news to his colleagues by e-mail (thus causing a certain commotion about what he dared to do…which will be undone at the end of the day by the actual CEO). In the field, we note that, even if the professional context somehow influences the levity of the game (learning issues, evaluation of one's colleagues, etc), as we have already seen, several of them, when they represent social games, allow leeway for "telling". Games demand a constant verbalization of who is who, who is doing what, how something should be done, why something has been done, and their goal and rules. Besides, playing together involves communication, whether during challenges, games with construction sets, or role-playing games. Pulling someone back into line or reminding him or her about the rules, the assessment of what is being played, as well as the subalterns' criticisms directed at management practices are all aspects that emerge in the game processes. What is often difficult to say can be put into words (in terms of bottom-up criticisms) in the game. Consequently, management use games in most cases to encourage discussions:

– Professional discussions of a technical kind: how to write an annual report, how to lead a subaltern to agree with what we want him or her to do, how to negotiate well, etc. The verbal aspect, in this case, concerns both the "best practices" and the participants' account about the problems they encounter, so that the issue can be shared and solved. Verbalization also concerns rules: what is right or appropriate to do or avoid doing, which has been explained in the part we dedicated to role.

– Discussions about a problem pointed out by management: Burn-out, psychosocial risks, discrimination, sexism, etc. Business theater, for example, is often used to this end. This can take place during seminars, in the shape of plays that an audience is invited (or not, which, in that case, invalidates this space for "telling") to respond to in order to start a debate about the problem raised by the business. This can also take place during group training when, at the end of the session, the participants may be encouraged to create their own sketch or compose their own song in order to quite rightly "tell" the others about their experience about the topic. It often happens that those who have organized the training, i.e. managers, are invited to the final performance, which entails two things: they directly receive their coworkers' message and they can assess how good the training provided is. If they do not go to see the theatrical creation of their employees, trainers have the task of reporting what has been said.

– Discussions aiming to strengthen team cohesion: the famous "teambuilding" process. During such games (Kapla blocks as well as theater, etc.), several objectives become clear: getting to know other people in the sense of being able to identify one's colleagues, finding out about each other in the game as well (a vehicle for relationships when people play together), and also seeing themselves function together and question how this can be done. This is the case for the "exercises in agreement": "What kind of link do we want to create between each other? Which of our behaviors toward each other can we agree on? Games are quite powerful tools to bring people together" (trainer). In some cases, when Kapla or Lego blocks are used, we can identify everyone's entrepreneurship (or lack thereof), adjustment strategies, bottlenecks in a team that has already been formed and competitive mindsets that have to be steered toward cooperation. All of this, which is unspoken, is destined to be verbalized during the game.

Some games will be clearly involved in the verbalization of what is unspoken. They may be used to create a real-life scenario: "Real-life scenarios must be created. I can't say to them 'Hi, we're gonna do some role-playing'. That's why I think that what I call 'the game of great replies' is important (…). I say to them: 'Tell me which replies, out of those you've heard, take your breath away' or what they hear during the day. Then I say to them 'Tell me what you'd feel like saying from the bottom of your heart'. Then we work on the replies" (trainer).

The game frame, in a work organization, literally offers a space for dialogue and allows leeway for speaking out. The fact that games not only involve fiction markers, which are often explicitly established ("Now John will play a colleague that feels bad"), but also require us to determine their related rules and goals entails that games, at least those that we have observed, are played by being spoken out.

3.4.5. *Promoting learning*

As we have been able to see, games are used in a twofold relation to knowledge:

– they are used to learn following a logic of training. In this case, they turn out to be both a medium and a context to do this: learning because of the game (from our reactions and actions when we play a role, for instance) but also creating a relaxed listening space good for learning;

– otherwise, we learn something from them while they take place, as is the case for reversal days and their TV adaptation. "The other" space of the game lets us change place, perspective, relationships, which is something that leads a director to learn – an "experience – about what he does not see or perceive about his company (especially because people would not dare to tell him about it, if we consider the verbatim accounts of the directors shown in the reality TV programs).

The list of educationalists and child psychologists who, in the 20th Century, discussed the relationships between games and learning is long. We have seen how Winnicott [WIN 71] regards them as a "potential space" and a "creative experience" favorable to the development of a child. Piaget, following a similar logic, says that games "allow us to adapt reality to the self, without constraints or sanctions" and "transforms the real world by incorporating it in a more or less absolute way into the needs of the self" [PIA 66, p. 46]. Vygotsky says about games that: "As in the focus of a magnifying glass, play contains all developmental tendencies in a condensed form and is itself a major source of development" (Vygotsky, conference held in 1933, published in 1966).

Viewing games as a principle of development "naturally" questions its relationship with learning, which has been determined by Western philosophy for a long time. We could mention Rabelais, who makes Gargantua play cards "to learn from them hundreds of little delights and novelties all of which derived from arithmetic" [RAB 34, p. 158)[17], or Rousseau in *Emile*, who highlights children's' games for instructive and developmental purposes[18]: "Who does not enjoy seeing a pretty child of this age, with his bright expression of serene content, and laughing, open countenance, playing at the most serious things (…)?" [ROU 62, p. 207). Henri Wallon will claim that games involve a logic of "functional progression" in the sense that they highlight a child's stages of development in relation to his or her acquisition of sensorimotor, symbolic and social functions [WAL 41].

As for adult learning, Ramirez and Squire [RAM 14] mention several types of learning, which they define "situated", linked to games – videogames in their study. Drawing on the studies carried out by Brown *et al.* [BRO 89] and relying on Vygotsky's research [VYG 78], they start from the cognitive sciences precept that the tools we use to organize our thoughts actually shape them. According to these authors, using videogames in a learning context would allow us to keep track of the progress made in terms of learning, to give feedback about this progress, to formulate with clarity the evaluation criteria, to motivate players to persevere, to encourage the (total) command of the game, to encourage individuals to take risks and explore new environments, to think about the rules of the game and their modification, to understand – thanks to the comparison with other

17 And he adds: "By such means [the game of cards] he developed a passion for the science of numbers, and every day after dinner and supper he would pass his time as pleasantly as he once had done with dice and playing cards.

18 While he actually disapproves of them when they are played by adults.

players – how there may be several different types of learning or to encourage cooperation.

What we notice about these possibilities provided by the game is that they are of three different kinds: one has to do with education (how to measure progress, determine the evaluation criteria and motivate: persistence and mastery), the second, quite explicit, concerns how playing teaches us about the game (taking risks, exploring, discussing rules in the game and in relation to it), while the third one is social (comparison and cooperation). Regarded in these terms, a game is an educational form or style assigned to learning processes: it teaches us something about this form and allows us to establish a relationship between the knowledge and practices of the game. According to Brougère, it is the features of what a game consists of that are vehicles for learning, which somehow lead to the ability to play in a figurative sense, to negotiate rules, and to take decisions on one's own within the context of the game, without forgetting the ability to "act in an uncertain context" [BRO 97, p. 50]. Here, we perceive the difficulty faced when measuring not only the scope of these learning processes, but also their duration or the fact that these abilities linked to the game can be, for example, transposed from the game context (do we develop an ability to read between the lines in the game or do we use an ability we already had? Would we run the same risks outside the game context? Would not a game eventually work only within its own boundaries, namely when and where it is played?). On the other hand, the contribution made by games underlined by Brougère is joined by what we may define as social skills, which structure our relationships with the other: this is what happens when we structure and agree on implicit matters and rules, when we take initiatives in relation to a given problem or issue, and when we behave in a flexible and adaptable way.

With the idea that what takes place within the game context is a learning process that has consequences "outside the frame", namely in "real life", we find again the notion of game as ritual [HAM 12, HAM 00]. With rituals, what happens within the ritual frame is assumed to take place in our world. Roberte Hamayon questions this aspect explicitly: "Are the imitative acts that characterize games necessarily of the same kind as the activities they are said to prepare us for? Could there be any continuity between the acts while there is, in principle, discontinuity between the fictional dimension of imitative playing and the actual context of what it is said to prepare us for in normal life?" [HAM 12, p. 142]. The social experience allowed by the game remains and maybe this is where we can find, more immediately but also more vaguely, the learning processes associated with it.

3.4.6. *The naturalization of rules and norms*

Bonenfant and Genvo [BON 14], as well as Brougère [BRO 05], highlight the framing process that games involve. First of all, games take place through their rules. Afterward, in our case, the games observed have been decided by management: both the object of the game and its form (challenge or role-playing game), where and when it takes place, as well as who plays. This aspect of the game corresponds to what Brougère says about it: "A game is not defined by a specific kind of experience, but by a tool used to produce it in a controlled way" [BRO 05, p. 102].

All evidence points to how games, in an educational context, represent a control tool. Games not only allow us to frame the debate, i.e. to mark the boundaries of the communications spaces by determining what they consist of (for example what is a problem or a priority), but they also enable us to point out which professional actions are appropriate and which are not. This framing process also has to do with behavior: when playing with other people, we involve our "faces" – as we have seen even outside a game context – colleague to colleague and not merely as players. In this case, it is the delicate identification and internalization of the margins of action that is entailed and shaped, since what can I allow myself to do as a professional-who-is-playing? What must be incorporated is not only the set of rules determining our social interactions and actions in relation to our colleagues in the workplace but also those rules that define how to apply and what to do with the "license" spaces (and their boundaries) that we are offered and whose goal consists of work or work relationships. What participants must take into account seems to be a double requirement to simultaneously take part in an action imposed by their management (firmly proposed, at best) and incorporate the message of supervised "freedom" conveyed through the use of "games". In this case, the process follows the logic of Goffmanian "performance", which pertains to methods as well as behaviors, dear to present-day management.

A tough exercise in which those who do not identify the boundaries with enough precision will fail, and which is identified as such, this expectation about internalizing boundaries becomes an implicit professional requirement. In the reversal days we observed, no one makes mistakes or dares to actually treat his or her boss like a colleague, leaving aside the ostensible nature of the game. Social rules are spread from one manager to another: "Interaction with another person involves a double contingency: the other's response is contingent upon one's own action, just as one's own action is contingent upon his. Furthermore, the contingent action is not an automatic response governed by mechanical laws, but a purposive action, directed by the actor's goals, and constrained by the rules of the social organization within which he is acting" [COL 69, p. 5]. Coleman adds that, in the context of the game and the interactions that take place in it, what is learnt is the "empirical regularities about the way other persons behave in particular situations" [COL 69, p. 5] according to rules and constraints they have to comply with. These regularities, which a player can thus identify, constitute what Coleman defines a "system of behavior" corresponding to a "system of rules" [COL 69]. In this respect, he agrees with what Bettelheim says about children's' games while stating that games, besides providing a space to "work through and master quite complex psychological difficulties of the past and present" and manage to deal with those of the moment, allow us to judge if the "various roles and forms of social interaction" [BET 87, p. 43] observed in the game can be suitable for us so that we can adopt them.

As for training, the educational frame redoubles the game frame (and the professional one) through the data required by the trainer-game leader. Role-playing games and their rules, as well as the purposes of the learning process (which are assumed to coincide since the latter are shaped by the former), are not questioned by the groups of participants. Participants start playing the game when they no longer discuss about what the game consists of or the blend of game, professional, and training frames. For example, in the game frame, they will comply with demands such as: "We're gonna play a 'flashfeedback'. You've got 20 seconds, a very short time, to help me re-evaluate myself about this: I'm a coworker of yours who works hard but who interrupts his client when he makes a comment on his work. You don't have the time to have a proper talk" (trainer). Then, in the educational frame, they will not question the trainer's advice during the debriefing: "giving him advice about what he's done makes the coworker dig in his heels even more.

What we have to give him is feedback about his future actions" (the same trainer).

During a training session for managers– advisors, as a theoretical supplement after a role-playing game, a trainer will allude to the experiment of the "electrified cage" tested on two rats in these terms: "A student will press a button to put the rat under a lot of stress at random intervals. The two rats have been suffering for the same amount of time but one of them can turn off the power on the bottom of the cage (for both cages). The second one's in great shape while the first one's dead while they have received the same shocks and felt the same pain. The question is: can people act on their environment? Those who do, well, they will see it in a positive light!" (trainer). The role of trainer, namely the practical manager of the training process, ensures, for example, that the principle of the experiment involving the animal's pain, as well as the fact that the rat in its cage could represent someone working in a company, is never called into question. The frame, in this case an educational one involving science (a scientific experiment), leads to the imposition of the validity of its terms as soon as it is formulated. It is performative.

During the training day, games will allow the performance as well as the verbalization of the knowledge resulting from theory, which participants are supposed to literally perform. Behaviors, non-verbal language and expressions are all scrutinized during the debriefings with the group: "The elements in these rules are not persons in the usual sense; they are actors-in-roles" [COL 69, p. 10]. What is played is backed up or called into question according to the goal consisting of how one's role should be played, which is quite a prominent aspect. Thus, a trainer will make managers work on how they should congratulate their employees by advising them to avoid any kind of negative congratulations such as: "there's nothing to say, you didn't really need to have your ass kicked, I can't see any screw-up here…". What trainers are working on, outside any kind of actual interpersonal context, is "how to make sure that congratulating someone is effective" (trainer).

In a different and more repetitive way, interpersonal norms can thus be worked on in group, with scenarios, and through the game: "what type of relationship do we want to establish between each other? Which way of behaving with each other can we agree on?"

Schild [SCH 68] and Coleman [COL 68] emphasize the significance of relationships and interactions, regarded as techniques in role-playing games (to lead someone else to agree to something, to agree on a performance, etc.). Schild adds that games are "amoral" since, regardless of whether people comply with the rules of the games, according to him no action "is intrinsically right or wrong. The only criterion is its effect on the player's score" [SCH 68, p. 98]. From this perspective, evaluation becomes essential and, for all practical purposes, the debriefings will be much longer than the games. Apart from the rules of the game which, according to Schild, consist of the goals to achieve, the resources that a player can use to reach them, and the "legitimate" actions performed to that end, a game, as such, establishes its own organization.

Therefore, we may think that games operate in the sense that they do a work of implementation, in acts, in accordance with normative expectations. Gilles Brougère underlines how "Nowadays, work conditions lead us to improve informal types of learning within organizations that are themselves required to be instructed" [BRO 05, p. 153]. In this case, there are different kinds of learning and their principle matches the one of socialization: an individual, through the exchanges, performances shown, and the control tools used by the structures he is immersed in, incorporates and claims codes, practices and behaviors for himself. All of these can be grasped by participating and cooperating rather than through the more distant orders and instructions given by the work organization to the individual. "Slides, they're useless!"

In this respect, games, by being introduced in a professional environment, are performative. They are dynamic and trivialize, as Bonenfant and Genvo – who have been previously mentioned – write, the principles of "accumulation, efficiency and productivity" [BON 14]. In relation to the cases studied, games banalize, for example, the temporal framework of the interactions, what is challenging, what must be prioritized and what has to be considered or left aside. By using games, order is less present and "reality is broken" [MCG 11] but this time in the sense that there is no reason to adopt a point of view while freely putting in perspective how work is conceived and organized. Who would criticize the idea of performance in games? Or the notion of competition? Or the concept of evaluation? Or the idea of setting a trap? These are much more delicate themes in the world of work. A

Marxist critique would, in this respect, refer to alienation or the organization of consent [BUR 79]. By making workers play and get involved in the game (by also avoiding those elements that make them accept or decide to take part), we make them assimilate and even embody their own modes of consent. In this respect, the developers' discussions about efficiency are instructive: "The more points you earn, the higher your level, and the higher your level, the more challenging work you unlock. This process is called 'leveling up'. The more challenging the work, the more motivated you are to do it, and the more points you earn...It's a virtuous circle of productivity" (McGonigal).

The strength of a social game such as a role-playing game lies in that, ever since childhood, we have learnt our "conduct rules" from others. However, as Jean-Marie Schaeffer underlines, "It goes without saying that the results of fictional training will never be as significant as those of 'real life' training" [SCH 99, p. 40]. This is an approach that this work will not deal with – this would lead us to abandon the study of the game frame and structure – and it also represents one of its boundaries.

Conclusion

From the Territory to the Map

Roberte Hamayon reminds us about simulation "models". This anthropologist consequently encourages us to question "the cognitive categories imparted by playing". The "reduced model" put forward by games: "(…) gives us a simplified representation of something complex, with an interpretative element that locates the action in a fictional frame so as to better choose what is relevant and dispose of its secondary aspects. The representation modeled is not a scaled-down reproduction. As a simplified and displaced interpretation, a reduced model determines a norm while also making it easily comprehensible and memorable" [HAM 12, pp. 159–160].

If we move from the model to the modeling process or, to use the image dear to Bateson [BAT 72], from the territory to the map, a selection as well as a reinterpretation of certain features assumed to be central to games takes place and has an effect. What is suggested to the players is a way of seeing the world. Games, as metaphors, distort the actual world that they are supposed to model. As is the case for any system of representation, certain parts of "what's real" are present within its framework and their arrangement, presentation, omission, or selection will propose an account, in relation to our topic, of what constitutes a kind of work to perform or hierarchical relationships.

From this perspective, we can – as we have done throughout this work – identify several of these transformations concerning a game frame related to a work context in those management games that can manifest themselves, as business theater does, in the shape of fiction. Each time, there is a "denoting" element, which is the "denoting" aspect of Bateson's theories:

the fake nip of the game "denoting" the real one without, however, being a non-bite or standing for a real bite. This gives rise to several pairs that can reveal the transformations at work in role-playing and simulation games:

– the logic of the character we have to play, defined by a mental state, an action, or a behavior (he does not want to have anything do with the project, he is very tired and depressed, he talks to his client rudely, etc.) versus the psychological aspect of an actual person, rather than a fictional character in a game;

– the subaltern played by a manager versus the social hierarchical relationships in the actual workplace;

– the time for playing a relationship, equivalent to the time of an interaction, versus the history of the relations among colleagues, which structure interpersonal relationships;

– following a similar logic, but in relation to the work organization, the business context described for the scene (motivating the decisions taken in this scene) versus the history of past decisions taken in the company and affecting the workers' choices. This clearly determines what is situated within the frame versus what is outside in historical as well as systemic dimensions (the "real" context where people do not decide just following the logic of interaction, but also by the whole of the context that surrounds it: other colleagues, preexisting representations, interpersonal or professional codes, etc.);

– the emotions played versus the actual emotions felt, such as weariness, the mental fatigue associated with burn-out, aggressiveness, etc.;

– the homogeneous groups (in terms of status and function) taking part in training sessions versus the heterogeneous groups (and points of view related to these positions) involved in work situations;

– consequently, the game partner versus the possible competitor in the company;

– the problems or priorities defined in the game (determined by management models) versus what actually represents a problem or a priority for the individuals in workplaces;

– generally, what is determined in the game (state) versus what crops up in a work situation, such as a problem or a priority, for instance (process);

– the logic of case (logic of choice or of selection) versus the simultaneity of events, facts and situations someone in the workplace has to deal with;

– the requirements of the game in terms of "a role to play" (the goals to achieve, which represent the rules of the game: for example having to pull an employee back into line) versus the immense freedom of choice enjoyed by people in "real life" (who may decide to avoid doing something, for instance) and their freedom to embody or not, according to the situation experienced, their role [SIM 08].

Games involving swapping places offer another series of oppositions between game frame and reference frame:

– game time versus working hours (weariness, routine, etc.): 2 h denoting – standing for – work in general in these games;

– consequently, what a director may know about work in the game versus what he may know about the work of his colleagues in his relationship with them: in the game, immersion stands for how a worker may communicate with his boss;

– closeness in context of the game versus the distance present in a work organization – many employees saying that it is the first time that they have met their boss: in the game, the closeness of the boss disguised as a worker denotes the boss's closeness with the other workers. This calls into question the forms of closeness, which, following this logic, can only become a reality when people share a similar hierarchical status;

– the worker embodied by the boss versus the actual worker: in the game, the boss-worker stands for the worker (we pretend that he is a worker);

– the boss-worker versus the actual boss: in the game, the boss-worker also stands for the boss (the double demand of the game is at work in this case: this individual is a boss and at the same time he is not a boss).

However, the game frame also simultaneously determines the boundaries of the frame in which certain actions taking place in the game do not stand for actions in a non-game context. The same can be said about the concept of evaluation, which, in the game, would not stand for the actual assessment of colleagues, or the boss in the game and his power to take decisions, which would not go over the game frame, and finally the reactions or remarks of the game partners, which would be invalid if we considered a relationship in a non-game context. Nonetheless, we have seen how the boundaries established between the frames could turn out to be porous, as is proved by how a boss, during the game, got involved in the work of an accountant, thus stepping outside the game frame. However, the lack of consequences is a recurrent theme that pertains to games and justifies their use in work organizations. The paradoxical dimension of the relation with consequences becomes even more relevant if we think that the game is expected to entail certain consequences in real life in terms of good behavior, adequate reactions, management techniques, assimilation of profession-related rules, etc.

It is this vagueness in relation to the actual distinction between game and work frames that makes gamification via role-playing and simulation games a complex topic. If fiction, frame, role and reversal are all mechanisms inherent to games, which are imported in non-game contexts, this specific gamification relying on social games raises certain problems that have to do more with social than game relationships. The very surrealist "this is not a boss" element gives us access – rather than to the absurd dimension of Magritte's work – to a system of interpretation concerning closeness, democracy, testing without fear of making a mistake without consequences, being exposed to someone else's perspective without being evaluated "for real", etc. In this respect, it keeps challenging/questioning the relationship between illusion and game, this "in-lusio" that has the same etymology, while also explicitly informing us about the types of present-day management discourses and ideals.

Bibliography

[ABT 70] ABT C., *Serious Games*, Viking Press, New York, 1970.

[AGI 00] AGIER M., *Anthropologie du carnaval*, Éditions Parenthèses/IRD, Marseille, 2000.

[ALV 07] ALVAREZ J., Du jeu vidéo au serious game, PhD Thesis, University of Toulouse, available at: http://ja.games.free.fr/These_SeriousGames/TheseSeriousGames.pdf, 2007.

[ALV 12] ALVAREZ J., DJAOUTI D., *Introduction au serious game*, Editions Questions théoriques, Paris, 2012.

[ANC 95] ANCELIN-SCHUTZENBERGER A., *Le jeu de rôle*, ESF, Paris, 1995.

[APP 01] APPADURAI A., *Après le colonialisme. Les conséquences culturelles de la globalization*, Payot, Paris, 2001.

[BAK 70] BAKHTINE M., *L'œuvre de François Rabelais et la culture populaire au Moyen Âge et sous la Renaissance,* Gallimard, Paris, 1970.

[BAT 72] BATESON G., *Steps to an ecology of mind: Collected essays in anthropology, psychiatry, evolution, and epistemology*, University of Chicago Press, Chicago, 1972.

[BAT 79] BATESON G., *Mind and Nature. A Necessary Unity*, Bantam Books, New York, 1979.

[BEL 06] BELMAS E., *Jouer autrefois. Essai sur le jeu dans la France moderne (XVI-XVIIIe siècles)*, Seyssel, Champ-Vallon, 2006.

[BER 94] BERCÉ Y.-M., *Fête et révolte. Des mentalités populaires du XVIe au XVIIIe siècle*, Hachette, Paris, 1994.

[BOL 99] BOLTANSKI L., CHIAPELLO E., *Le nouvel esprit du capitalisme*, Gallimard, Paris, 1999.

[BOO 68] BOOCOCK S.S., SCHILD E.O., *Simulation Games in Learning*, Sage Publications, LA, 1968.

[BOR 12] BORZAKIAN M. (ed.), "Les espaces ludiques", *Géographie et Cultures*, no. 82, L'Harmattan, Paris, 2012.

[BOU 94] BOURDIEU P., *Raisons pratiques*, Le Seuil, Paris, 1994.

[BRO 05] BROUGÈRE G., *Jouer/Apprendre*, Economica, Paris, 2005.

[BUR 79] BURAWOY M., *Manufacturing Consent: Changes in the Labor Process Under Monopoly CapitalismI*, University of Chicago Press, Chicago, 1979.

[BUR 85] BURAWOY M., *The Politics of Production: Factory Regime Under Capitalism and Socialism*, Verso Books, 1985.

[CAI 67] CAILLOIS R., *Les jeux et les hommes*, Gallimard, Paris, 1967.

[CAÏ 11] CAÏRA O., *Définir la fiction. Du roman au jeu d'échecs*, Editions de l'EHESS, Paris, 2011.

[CHA 07] CHAUVIER S., *Qu'est-ce qu'un jeu ?*, Vrin, Paris, 2007.

[CHA 10] CHATFIELD T., *Fun Inc.: Why Games are the 21st Century's Most Serious Business*, Random House, London, 2010.

[CRO 77] CROZIER M., FRIEDBERG E., *L'acteur et le système*, Le Seuil, Paris, 1977.

[DEJ 98] DEJOURS C., *Souffrance en France*, Le Seuil, Paris, 1998.

[DER 05] DERRIDA J., *Histoire du Mensonge, Prolégomènes*, L'Herne, Paris, 2005.

[DIB 07] DIBBELL J., *Play money: Or, how I quit my day job and made millions trading virtual loot*, Basic Books, New York, 2007.

[DUB 98] DUBAR C.T., *Sociologie des Professions*, Armand Colin, Paris, 1998.

[DUJ 12a] DUJARIER M.-A., Le management à Distance, Une Sociologie Clinique de L'Activité, Thesis, University Paris-Ouest, 13 December 2012.

[DUJ 12b] DUJARIER M.-A., *L'idéal au travail*, PUF, Paris, 2012.

[DUJ 15] DUJARIER M.-A., *Le management désincarné*, La Découverte, Paris, 2015.

[DUV 73] DUVIGNAUD J., *Fêtes et Civilisations*, Weber, Geneva, 1973.

[EDE 09] EDERY D., MOLLICK E., *Changing the Game*, FT Press, Upper Saddle River, 2009.

[FAB 92] FABRE D., *Le carnaval ou la fête à l'envers*, Gallimard, Paris, 1992.

[FAG 81] FAGEN R., *Animal Play Behavior*, Oxford University Press, Oxford, 1981.

[FRE 88] FREUD S., *L'inquiétante étrangeté et autres essais*, Gallimard, Paris, 1988.

[GAU 05] GAULEJAC DE V., *La société malade de la gestion*, Le Seuil, Paris, 2005.

[GEE 80] GEERTZ C., "Blurred Genres: the refiguration of social thought", *The American Scholar*, vol. 49 no. 2, pp. 165–179, 1980.

[GOF 61] GOFFMAN E., *Encounters. Two studies in the sociology of interaction*, Martino Publishing, 1961.

[GOF 63] GOFFMAN E., *Behavior in Public Places: Notes of the Social Organizations*, The Free Press, New York, 1963.

[GOF 67] GOFFMAN E., *Interaction Ritual: Essay on Face to Face Behavior*, Anchor Books, New York, 1967.

[GOF 74] GOFFMAN E., *Frame analysis: an Essay on the Organization of Experience*, Harvard University Press, Cambridge, 1974.

[GRO 96] GROOS K., *The Play of Animals*, Appleton, New York, 1896.

[GRO 08] GROOS K., *The Play of Man*, Appleton, New York, 1908.

[HAM 12] HAMAYON R., *Jouer. Une Etude Anthropologique*, La Découverte, Paris, 2012.

[HEE 71] HEERS J., *Fêtes, jeux et joutes dans les sociétés d'occident à la fin du Moyen-âge*, Paris, 1971.

[HEN 83] HENRIOT J., *Le jeu*, Synonyme-SOR, Paris, 1983.

[HEN 89] HENRIOT J., *Sous couleur de jouer. La métaphore ludique*, José Corti, Paris, 1989.

[HER 71] HERRON R.E., SUTTON-SMITH B. (eds), *Child's Play*, John Wiley and Sons, London, 1971.

[HUI 38] HUIZINGA J., *Homo Ludens. Essai sur la Fonction Sociale du Jeu*, Gallimard, Paris, 1938.

[IMM 80] IMMELMANN K., *Introduction to Ethology*, Plenum Press, New York, 1980.

[KAU 76] KAUFMAN A., FAURE R., LE GARFF A., *Les jeux d'entreprises*, PUF, Paris, 1976.

[LES 13] LESAVRE L., *Scènes de Management: Petit Manuel du Théâtre D'Entreprise*, PUG, Grenoble, 2013.

[LEV 62] LEVI-STRAUSS C., *La pensée sauvage*, Plon, Paris, 1962.

[LEV 74] LEVI-STRAUSS C., *Anthropologie Structurale*, Plon, Paris 1974.

[LIN 36] LINTON R., *The Study of Man: An Introduction*, D. Appleton-Century, New York, 1936.

[MCG 11] MCGONIGAL J., *Reality is Broken: Why Games Make Us Better and How They Can Change the World*, Penguin Books, London, 2011.

[MEA 34] MEAD G.H., *Mind, Self and Society*, University of Chicago Press, Chicago, 1934.

[MIL 79] MILLAR S., *La Psychologie du Jeu Chez les Enfants et les Animaux*, Petite Bibliothèque Payot, Paris, 1979.

[PIA 45] PIAGET J., *La Formation du Symbole Chez L'Enfant*, Delachaux et Niestlé, Neûchatel-Paris, 1945.

[PIA 66] PIAGET J., INHELDER B., *La Psychologie de L'Enfant*, PUF, Paris, 1966.

[PIE 88] PIETTE A., *Les Jeux de la Fête*, Publications de la Sorbonne, Paris, 1988.

[PRO 70] PROPP V., *Morphologie du Conte*, Le Seuil, Paris 1970.

[RAB 95] RABELAIS F., *La Vie Tres Horrificque du GRAND GARGANTUA, père de Pantagruel*, Le Seuil, Paris, 1995.

[ROU 62] ROUSSEAU J.-J., *Emile ou de l'éducation*, Garnier-Flammarion, Paris, 1762.

[ROY 05] ROY D., *Un Sociologue à L'Usine*, La Découverte, Paris, 2005.

[SAV 03] SAVIGNAC E., WASER A.-M., *Start-Up: Les Rois Ephémères*, Descartes et Cie, Paris, 2003.

[SCH 99] SCHAEFFER J.-M., *Pourquoi la Fiction*, Le Seuil, Paris, 1999.

[SHE 07] SHERMAN R., *Class Acts: Service and Inequality in Luxury Hotels*, University of California Press, Berkeley, 2007.

[SUI 05] SUITS B., *The Grasshopper: Games, Life and Utopia*, Broadview Press, Peterborough, 2005.

[SUT 97] SUTTON-SMITH B., *The Ambiguity of Play*, Harvard University Press, Cambridge, 1997.

[TAP 08] TAPSCOTT D., *Grown Up Digital: How the Net Generation is Changing Your World HC*, McGraw-Hill, New York, 2008.

[TUR 82] TURNER V.W., *From Ritual to Theatre: The Human Seriousness of Play*. Paj Publications, New York, 1982.

[TUR 90] TURNER V., *Le Phénomène Rituel, Structure et Contre-Structure*, PUF, Paris, 1990.

[WAL 14] WALZ S.P., DETERDING S. (eds), *The Gameful World*, MIT Press, Cambridge, 2014.

[WAL 41] WALLON H., *L'Evolution Psychologique de L'Enfant*, Armand Colin, Paris, 1941.

[WER 12] WERBACH K., HUNTER D., *For the Win: How Game Thinking can Revolutionize Your Business*, Wharton Digital Press, Pennsylvania, 2012.

[WIN 75] WINNICOTT D.W., *Jeu et Réalité. L'Espace Potentiel*, Gallimard, Paris, 1975.

[WIN 71] WINNICOTT D.W., *Playing and Reality*, Psychology Press, Oxford, 1971.

[WIN 88] WINNICOTT D.W., *Conversations Ordinaires*, Gallimard, Paris, 1988.

[WIT 53] WITTGENSTEIN L., *Investigations Philosophiques*, Gallimard, Paris, 1953.

[ZIC 13] ZICHERMANN G., JOSELIN L., *The Gamification Revolution: How Leaders Leverage Game Mechanics to Crush the Competition*, McGraw Hill, New York, 2013.

[ZIC 11] ZICHERMANN G., CUNNINGHAM C., *Gamification by Design: Implementing Game Mechanics in Web and Mobile Apps*., O'Reilly Media Inc., Sebastopol, 2011.

Articles/Chapters from books

[ABT 68] ABT C.C., "Games for learning" in SARANE S., BOOCOCK E.O., SCHILS, (eds), *Simulation Games in Learning*, Sage Publications, London, 1968.

[ALF 03] ALFEROFF C., KNIGHTS D., "We're all partying here: targets and games, or targets as games in call centre management", *Art and Aesthetics at Work*, Palgrave MacMillan, Manchester, 2003.

[ALL 15] ALLAL-CHERIF O., MAKHLOUF M., BAJARD A., "Les *serious games* au service de la gestion des ressources humaines: une cartographie dans les entreprises du CAC40", *Systèmes d'Information & Management*, vol. 19, no. 3, pp. 97–126, 2015.

[ANZ 11] ANZIEU D., "Les débuts du psychodrame en France (1946–1972)", *Revue de Psychothérapie Psychanalytique de Groupe*, no. 1, pp. 9–24, 2011.

[BAB 74] BABCOCK B.A., "The novel and carnival world", *Modern Language Notes*, vol. 89, no. 6, pp. 911–937, 1974.

[BAB 75] BABCOCK B.A., "A tolerated margin of mess: the trickster and his tales reconsidered", *Journal of the Folklore Institute*, vol. 11, no. 3 pp. 147–186, 1975.

[BAI 01] BAILLY R., "Le jeu dans l'œuvre de D.W. Winnicott", *Enfances & Psy*, vol. 3, no. 15, pp. 41–45, 2001.

[BAL 09] BALDRY C., JERRY H., "Welcome to the house of fun: Work space and social identity", in *Economic and Industrial Democracy*, Sage Publications, London, vol. 3, 2009.

[BAR 02] BARBUSSE B., "Sport et entreprise: des logiques convergentes ?", *L'Année Sociologique*, vol. 52, pp. 391–415, 2002.

[BAT 56] BATESON G., JACKSON D., DON H.J. *et al.*, "Toward a theory of schizophrenia", *Behavioral Science*, vol. 1, no. 4, pp. 251–254, 1956.

[BAT 71] BATESON G., "The message 'This is play'", in HERRON R.E., BRIAN S.-S. (eds), *Child's Play*, John Wiley and Sons, London, 1971.

[BAT 88] BATESON G., "Play and paradigm", *Play and Culture*, vol. 1, no. 1, pp. 20–27, 1988.

[BÉG 04] BÉGUIN P., CERF M., "Formes et enjeux de l'analyse de l'activité pour la conception des systèmes de travail", *Activités*, vol. 1, no. 1, pp. 54–71, 2004.

[BEN 47] BENVENISTE E., "Le jeu comme structure", *Deucalion*, 1947.

[BET 87] BETTELHEIM B., "The importance of play", *The Atlantic*, vol. 259, no. 3, pp. 35–46, 1987.

[BOG 14] BOGOST I., "Why gamification is bullshit" in WALZ S.P., DETERDING S. (eds), *The Gameful World*, MIT Press, Cambridge, 2014.

[BON 14] BONENFANT M., GENVO S., "Une approche située et critique du concept de gamification", *Sciences du Jeu*, available at: http://sciencesdujeu.univ-paris13.fr//index.php?id=7277, 2014.

[BOR 99] BORDELEAU P., "Technologies éducatives: déjà une longue histoire", *Les Cahiers du Millénaires*, available at: http://www.millenaire3.com/uploads/tx_ressm3/textes_bordeleau.Pdf, 1999.

[BOU 11] BOUTET M., "Un rendez-vous parmi d'autres. Ce que le jeu sur internet nous apprend du travail contemporain", available at: www.ethnographiques.org, no. 23, December 2011.

[BRO 87] BROMBERGER C., RAVIS-GIORDANI G., "Penser, agir et jouer avec le hasard", *Ethnologie Française*, nos. 2–3, pp. 129–136, 1987.

[BRO 89] BROWN J.S., COLLINS A., DUGUID P., "Situated cognition and the culture of learning", *Educational Researcher*, vol. 18, no. 1, pp. 32–42, 1989.

[BRO 97] BROUGÈRE G., "Jeu et objectifs pédagogiques: une approche comparative de l'éducation préscolaire", *Revue Française de Pédagogie*, no. 119, pp. 47–56, 1997.

[BRO 12] BROUGÈRE G., "Le jeu peut-il être sérieux? Revisiter Jouer/Apprendre en temps de serious game", *Australian Journal of French Studies*, vol. 49, no. 2, pp. 117–129, 2012.

[BRU 10] BRUNET M., RIFF J., "L'intérêt des jeux et rires pour la compréhension et la transformation des activités de travail", *Activités*, vol. 7, no. 2, pp. 7–2, available at http://activites. revues.org/2387, October 2010.

[BUR 08] BURAWOY M., "Le process de production comme jeu", *Tracés*, vol. 14, pp. 197–219, 2008.

[CAM 94] CAMPION M.A., CHERASKIN L., STEVENS M.J., "Career-related antecedents and outcomes of job rotation", *Academy of Management Journal*, vol. 37, no. 6, pp. 1518–1542, 1994.

[COE 05] COENEN-HUTHER J., *Heurs et Malheurs du Concept de Rôle Social*, Librairie Droz, 2005.

[COL 68] COLEMAN J., "Social processes and social simulation games", in SARANE S., BOOCOCK, SCHILD E.O. (eds), *Simulation Games in Learning*, Sage Publications, London, 1968.

[COL 69] COLEMAN J., "Games as vehicles for social theory", *American Behavioral Scientist*, vol. 12, pp. 2–6, 1969.

[COU 13] COURBET D., FOURQUET-COURBET M.-P., BERNARD F. *et al.*, "Communication persuasive et communication engageante pour la santé: Favoriser des comportements sains avec les médias, Internet et les serious games", in NATHALIE B. (ed.), *Publicité, Psychologie et Santé*, Editions In Press, Paris, 2013.

[CRO 75] CROZIER M., THOENIG J.-C., "La régulation des systèmes organisés complexes", *Revue Française de Sociologie*, vol. 16, pp. 3–32, 1975.

[CRO 09] CROOKALL D., THORNGATE W., "Acting, knowing, learning, simulating, gaming", *Simulation and Gaming*, vol. 40, no. 1, p. 8–26, 2009.

[DAN 07] DANIELLOU F., "Des fonctions de la simulation des situations de travail en ergonomie", *Activités Revue Électronique*, vol. 4, no. 2, pp. 77–83, 2007.

[DEJ 93] DEJOURS C., "Metis et Phronesis", *Education permanente*, no. 116, 1993.

[DEJ 12] DEJOURS C., GERNET I., "Travail, subjectivité et confiance", *Nouvelle Revue de Psychosociologie*, vol. 13, no. 1, pp. 75–91, 2012.

[DEL 09] DELCHAMBRE J.-P., "Le jeu créatif comme modalité de l'expérience dans une perspective pragmatique élargie", *Recherches Sociologiques et Anthropologiques*, pp. 15–38, available at http://rsa.revues.org/290, 2009.

[DES 06] DESCAMPS D., Ethologie du Jeu, Masters thesis, University Paris Descartes, 2006.

[DES 09] DESSORS D., "L'intelligence pratique", *Travailler*, no. 1, pp. 61–68, 2009.

[DET 11] DETERDING S., DIXON D., KHALED R. *et al.*, "From game design elements to gamefulness: defining gamification", *Proceedings of the 15th International Academic MindTrek Conference: Envisioning Future Media Environments*, pp. 9–15, 2011.

[DET 14] DETERDING S., "The ambiguity of games: histories and discourses of a gameful world" in WALZ S.P., DETERDING S. (eds), *The Gameful World*, MIT Press, Cambridge, 2014.

[DI 12] DI FILIPPO L., "Les notions de personnage-joueur et Roleplay pour l'étude de l'identité dans les MMORPG", *Revue ¿ Interrogations ?*, no. 15, available at http://www.revue-interrogations.org/Les-notions-de-personnage-joueur, December 2012.

[FLA 07] FLAMANT N., JEUDY-BALLINI M., "Le charme discret des entreprises", *Terrain*, no. 39, pp. 5–16, 2007.

[FOR 03] FORD R.C., FRANK S.M., JOHN W.N., "Questions and answers about fun at work", *Human Resource Planning* vol. 26, no. 4, pp. 18–33, 2003.

[GAR 01] GARRIGUES P., "Est-ce que les animaux jouent ?", *Enfances & Psy*, no. 15, pp. 11–17, 2001.

[GEE 80] GEERTZ C., "Jeu d'enfer. Notes sur le combat de coqs balinais", *Le Débat*, no. 7, pp. 86–146, 1980.

[GEN 06] GENVO S., "Les conditions de validité de l'immersion vidéoludique: pour une approche descriptive de la jouabilité", *Ludologique.com*, 2006.

[GEN 08] GENVO S., "Réflexions ludologiques", *Médiamorphoses*, no. 22, Paris, 2008.

[GEN 13] GENVO S., "Penser les phénomènes de ludicisation à partir de Jacques Henriot", *Sciences du jeu*, 1 October 2013.

[GIN 13] GIN CHOI Y., KWON J., KIM W., "Effects of attitudes vs experience of workplace fun on employee behaviors: Focused on Generation Y in the hospitality industry", *International Journal of Contemporary Hospitality Management*, vol. 25, no. 3, pp. 410–427, 2013.

[GIT 98] GITTLEMAN M., HORRIGAN M., JOYCE M., "'Flexible' workplace practices: evidence from a nationally representative survey", *Industrial & Labor Relations Review*, vol. 52, no. 1, pp. 99–115, 1998.

[GOF 57] GOFFMAN E., "The characteristics of total institutions", *Symposium on Preventive and Social Psychiatry*, pp. 77–78, 1957.

[GOF 81] GOFFMAN E., "A Reply to Denzin and Keller", *Contemporary Sociology*, vol. 10, pp. 60–68, 1981.

[GOG 11] GOGGIN J., "Playbour, farming and labour", *Ephemera: Theory and Politics in Organization*, vol. 11, no. 4, pp. 357–368, 2011.

[GRE 81] GREIMAS A.J., "Eléments pour une théorie de l'interprétation du récit mythique", *Communications*, vol. 8, pp. 34–65, 1981.

[GUI 83] GUILLAUME M., "Tu ne joueras point", *Réseaux*, vol. 1, no. 1, pp. 66–77, 1983.

[HAL 02] HALE FEINSTEIN A., MANN, S., CORSUN, D.L., "Charting the experiential territory: Clarifying definitions and uses of computer simulation, games, and role play", *Journal of Management Development*, vol. 21, no. 10, pp. 732–744, 2002.

[HAM 00] HAMAYON R., "Des usages de 'jeu' dans le vocabulaire rituel des peuples altaïques, *Jeux rituels*, *Études mongoles et sibériennes*: pp. 11–45, 2000.

[HUA 99] HUANG H.J., "Job rotation from the employees' point of view", *Research and Practice in Human Resource Management*, vol. 7, no. 1, pp. 75–85, 1999.

[JOR 05] JORGENSEN M., DAVIS K., KOTOWSKI S. *et al.*, "Characteristics of job rotation in the Midwest US manufacturing sector", *Ergonomics*, vol. 48, no. 15, pp. 1721–1733, 2005.

[KIN 00] KINNIE N., HUTCHINSON S., PURCELL J., "'Fun and surveillance': the paradox of high commitment management in call centres", *The International Journal of Human Resource Management*, vol. 11, no. 5, pp. 967–985, 2000.

[KÜC 05] KÜCKLICH J., "Precarious playbour: Modders and the digital games industry", *Fibre Culture*, vol. 5, no. 1, 2005.

[LAR 86] LARMAT JEAN L.R., LOUBET CHRISTIAN P.M., "Le Corps masqué: les changements de l'apparence", *Cahier du Centre D'études Médiévales de Nice*, no. 6, 1986.

[LEC 71] LECAVALIER G., "Les jeux de simulation dans l'enseignement de la sociologie", *Sociologie et sociétés*, vol. 3, no. 2, pp. 259–274, 1971.

[LE 89] LE GOFF J., "Rire au moyen-âge", *Les cahiers du centre de recherches historiques*, vol. 3, pp. 1–19, 1989.

[LE 13] LE LAY S., "Des pratiques ouvrières ludiques aux dispositifs managériaux ludistes: vers une instrumentalisation du jeu dans le travail", *La nouvelle revue du travail*, available at http://nrt.revues.org/702, 2013.

[LEI 88] LEIFER E., "Trials of involvement: evidence for local games", *Sociological forum*, vol. 3, pp. 499–524, 1988.

[LIC 08] LICOPPE C., "Logiques d'innovation, multiactivité et zapping au travail", *Hermès*, no. 50, pp. 171–178, 2008.

[MAC 84] MAC ALOON J.J., "Olympic games and the theory of spectacle in modern societies", in MAC A. (ed.), *Rite, Drama, Festival, Spectacle, Rehearsals Toward a Theory of Cultural Performances*, Institute for the Study of Human Issues, Philadelphia, 1984.

[MAR 85] MARTIN P., CARO T.M., "On the functions of play and its role in behavioral development", *Advances in the Study of Behavior*, vol. 15, pp. 59–103, 1985.

[MAR 05] MARTIN C.A., "From high maintenance to high productivity: What managers need to know about Generation Y", *Industrial and Commercial Training*, vol. 37, no. 1, pp. 39–44, 2005.

[MAU 10] MAUFFRET B., "Les différentes modalités du travestissement de l'homme en femme dans le carnaval de Cayenne", in NDAGANO B. (ed.), *Penser le Carnaval, Variations, Discours et Représentations*, Karthala, Paris, 2010.

[MOL 01] MOLLICK E.R., ROTHBARD N., "Mandatory fun: consent, gamification and the impact of games at work", *The Wharton School Research Paper Series*, 2014, 2001.

[MOL 14] MOLLICK E., WERBACH K., "Gamification and the enterprise", in WALZ S.P., DETERDING S. (eds), *The Gameful World*, MIT Press, Cambridge, 2014.

[MON 01] MONJARET A., "La fête, une pratique extra-professionnelle sur les lieux du travail", *Revue Cités*, no. 8, pp. 87–100, 2001.

[NAY 01] NAYYAR S., "Inside the mind of Gen Y", *American Demographics*, available at: http://adage.com/article/american-demographics/inside-mind-gen-y/43091/, vol. 23, no. 9, p. 6, 2001.

[NEL 12] NELSON M.J., "Soviet and American precursors to the gamification of work", *Proceeding of the 16th International Academic MindTrek Conference*, pp. 23–26, 2012.

[NEW 12] NEWELL L.A., "Happiness at the house of mouse: how disney negotiates to create the happiest place on Earth", *Pepp. Disp. Resol. LJ*, vol. 12, p. 415, 2012.

[ONO 14] ONOFREI G., SIMON S., "Simulation games in operations management: the importance of immediate post game analysis", *Global Management Journal*, vol. 6, no. 1, pp. 61–64, 2014.

[PAS 06] PASTRÉ P., MAYEN P., VERGNAUD G., "La didactique professionnelle", *Revue française de Pédagogie. Recherches en Education*, no. 154, pp. 145–198, 2006.

[PAT 05] PATIN B., "Le jeu de rôles: pratique de formation pour un public d'adultes", *Les Cahiers Internationaux De Psychologie Sociale*, pp. 163–178, 2005.

[PAU 01] PAUL P., "Getting inside gen Y", *American Demographics*, vol. 23, no. 9, pp. 42–49, 2001.

[PÉC 01] PÉCHARMAN M., "Le divertissement selon Pascal ou la fiction de l'immortalité", *Cités*, no. 7, pp. 13–19, 2001.

[PET 84] PETERSEN A.F., GARRIGUES P., ROQUEFEUIL L.G., "Jeu et activité créatrice: le jeu en tant que résolution de problème chez l'animal et l'enfant", in GUILLEMOT J., MYQUEL M., SOULAYROL R. *et al.* (eds), *L'Enfant,* Expansion Scientifique Française, Paris, 1984.

[PER 67] PERROT C.-H., "Be di murua: un rituel d'inversion sociale dans le royaume agni de l'Indénié", *Cahiers D'Études Africaines*, vol. 7, no. 27, pp. 434–443, 1967.

[PIC 08] PICARD M., "Video games and their relationship with other media" in *The Video Game Explosion: A History from Pong to Playstation an Beyond*, Greenwood Press, Westport, 2008.

[PIC 12] PICHAULT F., PLEYERS M., "Pour en finir avec la génération Y... Enquête sur une représentation managériale", *Annales des Mines-Gérer et Comprendre*, ESKA, pp. 39–54, Paris, 2012.

[PIE 88] PIETTE A., "L'intervalle festif. Hypothèses théoriques et problématique de recherche", *Cahiers Internationaux de Sociologie*, vol. 85, pp. 325–342, 1988.

[PIE 92] PIETTE A., "Play, reality and fiction: toward a theoretical and methodological approach to the festival framework", *Qualitative Sociology*, vol. 15, no. 1, pp. 37–52, 1992.

[PIE 97] PIETTE A., "Pour une anthropologie comparée des rituels contemporains. Rencontre avec les batesoniens", *Terrain*, vol. 29, pp. 139–150, 1997.

[PIE 05] PIETTE A., "Fête, spectacle, cérémonie: des jeux de cadres", *Revue Hermès*, vol. 43, pp. 39–46, 2005.

[PRA 09] PRALONG J., "La 'génération Y' au travail: un péril jeune?", *Actes du XXème Congrès de L'AGRH*, Toulouse, 2009.

[QUE 03] QUERREC R., BUCHE C., MAFFRE E. *et al.*, "SécuRéVi: virtual environments for fire-fighting training", *5th Virtual Reality International Conference (VRIC'03)*, pp. 169–175, 2003.

[RAC 13] RACZKOWSKI F., "It's all fun and games… A history of ideas concerning gamification", *Proceedings of DiGRA, DeFragging Game Studies*, pp. 44–54, 2013.

[RAM 14] RAMIREZ D., SQUIRE K., "Gamification and learning", in WALZ S.P., DETERDING S. (eds), *The Gameful World*, MIT Press, Cambridge, 2014.

[REE 09] REEVES B., READ J.L., "Total engagement", in REEVES B., READ J.L. (eds), *Using Games and Virtual Worlds to Change the Way People Work and Businesses Compete*, Harvard Business Review Press, Boston, 2009.

[ROB 60] ROY R., "Banana time; Job satisfaction and informal organization", *Human Organization*, vol. 18, pp. 158–168, 1960.

[ROB 10] ROBERTSON M., "Can't play, won't play", available at http://www.hideandseek.net/2010/10/06/cant-play-wont-play/, Hideandseek, New York, 2010.

[RUI 13] RUIZALBA R., JOSÉ L., FERNANDO N.L. *et al.*, "Gamificación como estrategia de marketing interno, (Spanish)", *Intangible Capital,* vol. 9, no. 4, pp. 1113–1144, 2013.

[SAL 89] SALAMONE F.A., "Anthropology of play: a bibliography", *Play and Culture*, vol. 2, no. 2, pp. 158–181, May 1989.

[SAL 08] SALGADO M., "Formation au management: les vertus du théâtre en formation continue et en formation initiale", *14° Colloque de la Recherche des IUT: Développement Durable*, Lyon, 2008.

[SAV 13] SAVIGNAC E., "Carnaval d'entreprise et spectacularisation de l'ordre managérial", *Journal des Anthropologues*, no. 1, pp. 385–407, 2013.

[SAV 11] SAVIGNAC E., "Il était une fois... mon patron à ma place. Renversement des hiérarchies et morphologie du conte", *Jeunes et Médias*, vol. 1, pp. 79–93, Paris, 2011

[SCH 68] SCHILD E.O., "Interaction in games", in SARANE S., BOOCOCK, SCHILS E.O. (eds), *Simulation Games in Learning*, Sage Publications, London, 1968.

[SCH 88] SCHECHNER R., "Playing", *Play and Culture*, vol. 1, no. 1, pp. 1–2, 1988.

[SCH 91] SCHWARTZMAN H.B., "Imagining play", *Play and Culture*, vol. 4, no. 3, pp. 214–222, 1991.

[SIM 08] SIMPSON B., CARROLL B., "Reviewing 'role' in processes of identity construction", *Organization*, vol. 15, no. 1, pp. 29–50, 2008.

[STA 89] STAR S.L., GRIESEMER J.R., "Institutional ecology, 'translations' and boundary objects: Amateurs and professionals in Berkeley's Museum of Vertebrate Zoology", *Social Studies of Science*, vol. 19, no. 3, pp. 387–420, 1989.

[STE 04] STEICHEN R., "Plus de ruse que de raison. Du point de vue de l'anthropologie clinique", in LATOUCHE S. *et al.* (eds), *Les Raisons de la Ruse, Une Perspective Anthrpologique et Psychanalytique*, La découverte/MAUSS, Paris, 2004.

[SWE 01] SWEDBERG R., "Sociology and game theory: contemporary and historical perspectives", *Theory and Society*, vol. 30, pp. 301–335, 2001.

[THI 93] THIAGARAJAN S., "How to maximize transfer form simulation games through systematic debriefing", in PERCIVAL F., LODGE S., SAUNDERS D. (eds.), *The Simulation and Gaming Yearbook, 1993: Developing Transferable Skills in Education and Training*, Kogan, London, 1993.

[TOU 13] TOUZET L., "Carnaval in Rio: Dionysian Drama in an Industrializing Society", in TURNER V., SCHECHNER R. (eds), *The Anthropology of Performance*, PAJ Publications, New York, 2013.

[VAN 91] VAN MAANEN J., "The smile factory: work at Disneyland", in FROST P.J. *et al.* (eds), *Reframing Organisational Culture*, SAGE, London, 1991.

[VET 05] VETTIN J., "Human laughter, social play, and play vocalizations of non-human primates: an evolutionary approach", *Behaviour*, vol. 142, no. 2, pp. 217–240, 2005.

[VOL 91] VOLKWEIN KARIN A.E., "Play as a path for liberation: a Marcusean perspective", *Play and Culture*, vol. 4, no. 4, pp. 359–370, 1991.

[VYG 66] VYGOTSKI L.S., "Le rôle du jeu dans le développement de l'enfant", *Voprosi Psikhologii*, available at: http://www.ich-sciences.de/fileadmin/texte/leroledujeudans.pdf, no. 6, pp. 62–72, 1966.

[WEN 96] WENDLING T., "Compte-rendu du Désordre du jeu", *L'Homme*, vol. 36, no. 139, pp. 156–157, 1996.

[WEN 10] WENDLING T., "Une joute intellectuelle au détriment du jeu ? Claude Lévi-Strauss vs Roger Caillois (1954–1974)", *Ethnologies*, vol. 32, pp. 29–49, 2010.

[WIN 80] WINNYKAMEN F., CORROYERD., "Relations sociales entre enfants de 2; 6 à 3; 0. Rôle de l'imitation et étude des interactions", *Bulletin de Psychologie*, vol. 348, pp. 77–84, 1980.

[YOU 13] YOUNG G.C., JUNEHEE K., WANSOO K., "Effects of attitudes vs experience of workplace fun on employee behaviors: Focused on Generation Y in the hospitality industry", *International Journal of Contemporary Hospitality Management*, vol. 25, no. 3, pp. 410–427, 2013.

[ZAC 05] ZACCAÏ-REYNERS N., "Fiction et typification. Contribution à une approche théorique de la transmission de l'expérience", *Méthodos*, available at http://methodos.revues.org/378, 2005.

[ZAC 06] ZACCAÏ-REYNERS N., "Jouer pour penser ? Expérience ludique et réflexivité", in ENRIQUEZ E., HARICTE C., SPURK J. (eds), *Désir de Penser, Peur de Penser*, Parangon, Paris, 2006.

Filmography

Jouer pour grandir, Fayda Winnykamen, 2001, GUILGAMESH, La cinquième, RTBF, To Do Today.

La gueule de l'emploi, Didier Cros, 2011, Zadig Production.

La mise à mort du travail, Jean-Robert Viallet, 2009, Yami Production.

L'univers social des macaques, Bernard Thierry, 1990, CNRS Images.

Vis ma vie, Reservoir Prod.

"Directeur de parc" (diffusion: 19/09/2006), "Restauration" (17/10/2006), "Jean-Claude Debard, patron de Hyundaï" (05/12/2006), "Godiva" (20/02/2007), "Casino" (13/03/2007) and "Thalasso" (17/04/2007).

Patron incognito, Endemol Production.

Episodes de 2012: Jean-Claude Puerto (UCAR), Jean-Manuel Bajen (Bajen S.A.), Guillaume Richard (O2) et Nicolas Riché (Columbus café).

Episode from 2014: Rodolphe Wallgren (Memphis café).

Episode from 2015: Karine Cohen (Coursier.com).

Index

A, B, C

D, E, F

G, H, I

J, L, M

N, P, R

S, T, U

Other titles from

iSTE

in

Innovation, Entrepreneurship and Management

2016

BARBAROUX Pierre, ATTOUR Amel, SCHENK Eric
Knowledge Management and Innovation (Smart Innovation Set – Volume 6)

BEN BOUHENI Faten, AMMI Chantal, LEVY Aldo
Banking Governance, Performance And Risk-Taking: Conventional Banks Vs Islamic Banks

BOUTILLIER Sophie, CARRE Denis, LEVRATTO Nadine
Entrepreneurial Ecosystems (Smart Innovation Set – Volume 2)

BOUTILLIER Sophie, UZUNIDIS Dimitri
The Entrepreneur (Smart Innovation Set – Volume 8)

BOUVARD Patricia, SUZANNE Hervé
Collective Intelligence Development in Business

GALLAUD Delphine, LAPERCHE Blandine
Circular Economy, Industrial Ecology and Short Supply Chains (Smart Innovation Set – Volume 4)

GUERRIER Claudine
Security and Privacy in the Digital Era (Innovation and Technology Set – Volume 1)

MEGHOUAR HICHAM
Corporate Takeover Targets

MONINO Jean-Louis, SEDKAOUI Soraya
Big Data, Open Data and Data Development
(Smart Innovation Set – Volume 3)

MOREL Laure, LE ROUX Serge
Fab Labs: Innovative User
(Smart Innovation Set – Volume 5)

PICARD Fabienne, TANGUY Corinne
Innovations and Techno-ecological Transition (Smart Innovation Set – Volume 7)

2015

CASADELLA Vanessa, LIU Zeting, DIMITRI Uzunidis
Innovation Capabilities and Economic Development in Open Economies
(Smart Innovation Set – Volume 1)

CORSI Patrick, MORIN Dominique
Sequencing Apple's DNA

CORSI Patrick, NEAU Erwan
Innovation Capability Maturity Model

FAIVRE-TAVIGNOT Bénédicte
Social Business and Base of the Pyramid

GODÉ Cécile
Team Coordination in Extreme Environments

MAILLARD Pierre
Competitive Quality and Innovation

MASSOTTE Pierre, CORSI Patrick
Operationalizing Sustainability

MASSOTTE Pierre, CORSI Patrick
Sustainability Calling

2014

DUBÉ Jean, LEGROS Diègo
Spatial Econometrics Using Microdata

LESCA Humbert, LESCA Nicolas
Strategic Decisions and Weak Signals

2013

HABART-CORLOSQUET Marine, JANSSEN Jacques, MANCA Raimondo
VaR Methodology for Non-Gaussian Finance

2012

DAL PONT Jean-Pierre
Process Engineering and Industrial Management

MAILLARD Pierre
Competitive Quality Strategies

POMEROL Jean-Charles
Decision-Making and Action

SZYLAR Christian
UCITS Handbook

2011

LESCA Nicolas
Environmental Scanning and Sustainable Development

LESCA Nicolas, LESCA Humbert
Weak Signals for Strategic Intelligence: Anticipation Tool for Managers

MERCIER-LAURENT Eunika
Innovation Ecosystems

2010

SZYLAR Christian
Risk Management under UCITS III/IV

2009

COHEN Corine
Business Intelligence

ZANINETTI Jean-Marc

Sustainable Development in the USA

2008

CORSI Patrick, DULIEU Mike
The Marketing of Technology Intensive Products and Services

DZEVER Sam, JAUSSAUD Jacques, ANDREOSSO Bernadette
Evolving Corporate Structures and Cultures in Asia / Impact of Globalization

2007

AMMI Chantal
Global Consumer Behavior

2006

BOUGHZALA Imed, ERMINE Jean-Louis
Trends in Enterprise Knowledge Management

CORSI Patrick *et al.*
Innovation Engineering: the Power of Intangible Networks

Printed and bound by CPI Group (UK) Ltd, Croydon, CR0 4YY